WALCH PUBLISHING

20.00

Daily Warm-Ups

GENERAL MATH

Brian Pressley

Level I

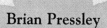

D1287863

Certified Chain of Custody
Promoting Sustainable
Forest Management
www.sfiprogram.org

SUSTAINABLE
FORESTRY
INITIATIVE

SGS-SFI/COC-US09/5501

The classroom teacher may reproduce materials in this book for classroom use only.

The reproduction of any part for an entire school or school system is strictly prohibited.

No part of this publication may be transmitted, stored, or recorded in any form

without written permission from the publisher.

1   2   3   4   5   6   7   8   9   10

ISBN 0-8251-6065-0

Copyright © 2006

J. Weston Walch, Publisher

P.O. Box 658 • Portland, Maine 04104-0658

walch.com

Printed in the United States of America

The *Daily Warm-Ups* series is a wonderful way to turn extra classroom minutes into valuable learning time. The 180 quick activities—one for each day of the school year—practice general math skills. These daily activities may be used at the very beginning of class to get students into learning mode, near the end of class to make good educational use of that transitional time, in the middle of class to shift gears between lessons—or whenever else you have minutes that
now go unused.

Daily Warm-Ups are easy-to-use reproducibles—simply photocopy the day's activity and distribute it. Or make a transparency of the activity and project it on the board. You may want to use the activities for extra-credit points or as a check on the math skills that are built and acquired
over time.

However you choose to use them, *Daily Warm-Ups* are a convenient and useful supplement to your regular lesson plans. Make every minute of your class time count!

Integers

Integers are all the whole numbers and their opposites. For example, 3 and –3 are both integers. The positive integers may be written with a plus sign in front of them (such as +1, +2, +3), or without (such as 1, 2, 3). The negative integers are always marked with a negative sign (such as –3, –2, –1). The set of integers also includes zero, which is neither positive nor negative. The three periods used before and after a set of integers, called ellipses, indicate that the set of integers continues indefinitely in both directions.

Answer the following.

1. What integer best describes a drop of 20°?

2. What integer best describes an increase of 20°?

3. What integer best describes a location 300 feet above sea level?

4. What integer best describes a location 300 feet below sea level?

5. Draw a number line that includes the integers 1, 9, –5, 8, –2, and 2.

6. Draw a number line that includes the integers 0, 25, 50, –100, 150, –150, and –75.

1

© 2006 Walch Publishing

Absolute Value

Absolute value is a measure of how far a number is from zero on the number line. The symbol for absolute value is a pair of parallel, vertical lines, one to the left and one to the right of a number, as shown here: |5| The absolute value of any positive number is just the number itself (for example, |5| = 5). The absolute value of any negative number is the number without the negative sign (for example, |–5| = 5). Absolute value is useful when you want to know an amount of change. For example, if the temperature in a place increases from 50°F to 70°F, the temperature change is clearly 20°F. But what if the temperature drops from 70°F to 50°F? Is the temperature change –20°F? No. The change is an absolute value. |20| = |–20| = 20.

Find the absolute value of each number below.

1. 8

2. 56

3. –22

4. – 6

5. 100

6. –524

7. 66

8. –30

9. 99

10. –5

2

© 2006 Walch Publishing

Comparing Integers

Consider the following set of integers: 4, 18, –3, 5, 12, 7, –8, 9, 21, 0, –2. They cover a broad range of numbers and include positive numbers, negative numbers, and zero. These numbers are also out of order. In order, they would appear as –8, –3, –2, 0, 4, 5, 7, 9, 12, 18, 21. We generally list integers from smallest to largest. It is important to remember that the greater the absolute value of a negative number, the "smaller" that number is. This means that we start our list with the negative numbers that have the largest absolute value and go by decreasing absolute value until we reach zero. Then we list positive integers by increasing absolute value.

Order each set of numbers below from smallest to largest.

1. 1, 1, 2, 2, 3, 3, 7, 8, 1, 3, 2, 5, 4, 6, 6, 1, 2, 1

2. 17, 18, 21, 20, 17, 17, 19, 19, 18, 20, 21, 21, 17, 18, 19, 17

3. 200, –25, 175, –100, –300, –75, 150, 50, 0

4. –12, –21, –30, –8, –16, –2, –33, –7

5. 1000, – 45, 654, 2, –306, 55, 921, –341

3

© 2006 Walch Publishing

Adding Integers

Although you are already familiar with adding numbers, the way we look at the addition of integers is little bit different. An integer's absolute value depends on its distance from zero on the number line. If you take 5 and add a positive number such as 3 to it, you move three numbers to the right on the number line. This would put you on the number 8, and as you know, 5 + 3 = 8. If however, you take 5 and add –3 to it, you move three spaces to the left. This would put you on the number 2, and as you know, 5 – 3 = 2. However, you didn't subtract; you added a negative. The equation should be written as 5 + (–3) = 2. If you had started with –8 and added –3 to it, you would move three spaces to the left on the number line. This would put you on –11. The equation should be written as –8 + (–3) = –11.

4

Solve the following addition problems.

1. 10 + 7
2. 10 + (–7)
3. –15 + 6
4. –12 + (– 4)

5. 25 + 60
6. 25 + (– 60)
7. –250 + 55
8. –250 + (–55)

© 2006 Walch Publishing

Subtracting Integers

Although you are already familiar with subtracting numbers, the subtraction of integers is a little different. An integer's absolute value depends on its distance from zero on the number line. If you take 10 and subtract a positive number such as 3 from it, you move three numbers to the left on the number line. This would put you on the number 7, and as you know, $10 - 3 = 7$. If, however, you take 10 and subtract -3 from it, you move three spaces to the right. This would put you on the number 13, and as you know, $10 + 3 = 13$.

However, you didn't add; you subtracted a negative. The equation should be written as $10 - (-3) = 13$. If you had started with -12 and subtracted -3 from it, you would move three spaces to the right on the number line. This would put you on -9. The equation should be written as $-12 - (-3) = -9$.

Solve the following subtraction problems.

1. $10 - 7$

2. $10 - (-7)$

3. $-15 - 6$

4. $-12 - (-4)$

5. $25 - 60$

6. $25 - (-60)$

7. $-250 - 55$

8. $-250 - (-55)$

5

© 2006 Walch Publishing

Multiplying Integers

Multiplication is repeated addition. For example, 4 × 8 = 8 + 8 + 8 + 8 = 32. This tells us that four 8s add up to 32. Multiplying a positive integer by a positive integer will move the value of the product toward the right on the number line. For example, 8 × 3 = 24. Multiplying a positive integer by a negative integer will move the total product to the left on the number line. For example, 8 × (–3) = –24. Multiplying a negative integer by a negative integer will move the value of the product to the right on the number line. For example, –10 × (– 4) = 40.

Remember, when two numbers of the same sign are multiplied, the product is positive. When two numbers of opposite signs are multiplied, the product is negative.

6

Solve the following multiplication problems.

1. 4 × 16

2. 5 × –8

3. –10 × –3

4. 9 × 11

5. 11 × –16

6. –25 × 25

7. 525 × 10

8. 5 × –2500

© 2006 Walch Publishing

Dividing Integers

Division is repeated subtraction. For example, 28 ÷ 7 = 28 – 7 – 7 – 7 – 7 = 0. This tells us that you can subtract 7 out of 28 four times. Division of integers follows the same basic rules as multiplication of integers. Dividing a positive integer by a positive number will produce a positive quotient. For example, 21 ÷ 7 = 3. Dividing a positive integer by a negative number will produce a negative quotient. For example, 24 ÷ (–3) = –8. Dividing a negative integer by a negative number will produce a positive quotient. For example, –20 ÷ (– 4) = 5.

When two numbers of the same sign are divided, the quotient is positive. When two numbers of opposite signs are divided, the quotient is negative.

Solve the following division problems.

1. 800 ÷ 20

2. 800 ÷ –20

3. –800 ÷ –20

4. 54 ÷ 9

5. 66 ÷ –11

6. –225 ÷ –25

7. 2004 ÷ 6

8. 2030 ÷ –35

7

© 2006 Walch Publishing

Prime Factors

When two numbers are multiplied together, those two numbers are called **factors**. A **prime number** is a whole number that is greater than 1 and that has only two unique factors, the number itself and 1. For example, 5 is a prime number because its only factors are 1 and 5. 6 is not a prime number because its factors are 1 and 6, as well as 2 and 3. The number 6 is a composite number. A **composite number** is a whole number greater than 1 that has more than two factors.

Determine if each of the following numbers is prime, composite, or neither. If the number is composite, list all the factors for that number.

8

1. 5

2. 9

3. 12

4. 0

5. 35

6. 16

7. 8

8. 50

9. 19

10. 0.5

11. 1

12. 10

13. 15

14. 17

© 2006 Walch Publishing

Powers and Exponents

A number such as 3^2 has a base number (3) and an exponent (2). A number that is shown as a combination of a base and an exponent is called a **power.** Powers are a shorter way of writing some longer expressions.

For example: $5 \times 5 = 5^2$
$6 \times 6 \times 6 \times 6 = 6^4$
$a \cdot a \cdot a = a^3$
$X \cdot X \cdot X \cdot Y \cdot Y \cdot Y \cdot Y = X^3 Y^4$

Show the following in exponent form.

1. $9 \times 9 \times 9 \times 9$

2. $h \cdot h \cdot h \cdot h \cdot h \cdot h \cdot h$

3. $m \cdot m \cdot n \cdot n \cdot n \cdot n \cdot n$

4. $10 \times 10 \times 10$

Answer the following.

5. Write the prime factors of 36 in exponent form.

6. What is the value of the expression 5^4?

7. What is a^3 if $a = 4$?

8. Which of the following has the largest value: 2^5, 3^4, 4^5, or 5^4?

9

© 2006 Walch Publishing

Square Roots

A factor that you multiply by itself to get a number is called the **square root** of that number. For example, the square root of 36 is 6 because 6 × 6 equals 36. A **perfect square** is a number whose square root is a whole number. 64, 100, and 9 are all examples of perfect squares. The mathematical symbol for a square root is called a radical and looks like this: $\sqrt{}$. Used in an example, $\sqrt{81}$ = 9.

Some square roots are not whole numbers. For example, the square root of 12 is approximately 3.46. In a case such as this, you may need to use a calculator to find the square root of a number.

Find the square root of each number below.

1. 4
2. 121
3. 16
4. 49
5. 36

6. 50
7. 85
8. 5000
9. 1,000,000
10. 268.96

10

Daily Warm-Ups: General Math

© 2006 Walch Publishing

Order of Operations

When faced with a long math statement containing many operations such as $20^2 \div 4 \times 10 + (15 \div 3) - 28 \div 7$, you might wonder where you should start. There are four basic rules that should help you perform this calculation in the correct order.

For example, let's simplify the expression mentioned above.

First, simplify inside parentheses: $20^2 \div 4 \times 10 + (5) - 28 \div 7$

Then simplify powers: $400 \div 4 \times 10 + 5 - 28 \div 7$

Then multiply and divide from left to right: $1000 + 5 - 4$

Then add and subtract from left to right: $1000 + 5 - 4 = 1001$

Simplify each problem below using the correct order of operations.

1. $100 + 4 - 25 \div 5 \times 6$

2. $75 + 8 \times 6 - 21 \div 3$

3. $((65 \times 5 - 5) + 20) \div 4 + 30$

4. $25^2 \div 5 \times 12 + (75 \div 3 + 4^2) - 121 \div 11$

11

© 2006 Walch Publishing

Rational Numbers

A **rational number** is a number that can be expressed as a fraction. Fractions, repeating decimals, terminating decimals, and integers are all rational numbers. More technically, a rational number is a number that can be expressed in the form $\frac{x}{y}$ where x and y are integers and y cannot be zero. The Venn diagram below shows the relationship between whole numbers, integers, and rational numbers.

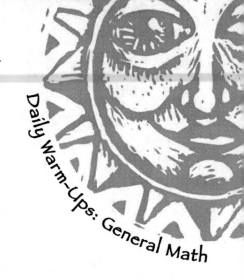

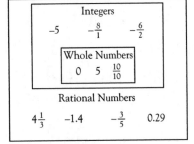

Integers

$-5 \qquad -\frac{8}{1} \qquad -\frac{6}{2}$

Whole Numbers

$0 \quad 5 \quad \frac{10}{10}$

Rational Numbers

$4\frac{1}{3} \qquad -1.4 \qquad -\frac{3}{5} \qquad 0.29$

Determine what group each number belongs to using the Venn diagram above. Put each number into the smallest group possible.

12

1. 4

2. −7

3. $\frac{1}{5}$

4. 100

5. 0.89

6. − 6.5

7. $-\frac{7}{1}$

8. $\frac{1}{3}$

© 2006 Walch Publishing

Irrational Numbers

An **irrational number** is a number that cannot be expressed as a fraction. Any decimals that are not terminating and do not repeat are irrational numbers. More technically, an irrational number is a number that cannot be expressed in the form $\frac{x}{y}$ where x and y are integers and y is not zero. The square root of 2 is a common example. If you take the square root of 2 on a calculator, you could possibly see 1.414213562373095 . . ., and with a calculator capable of showing more decimal places you would see that the numbers do not repeat. Actually, the square roots of all the numbers that are not perfect squares are irrational numbers.

Identify each number below as rational or irrational.

1. $\sqrt{4}$

2. $\sqrt{14}$

3. $\sqrt{\frac{27}{3}}$

4. $\frac{2}{3}$

5. $\sqrt{256}$

6. $\sqrt{500}$

7. $\sqrt{5.5}$

8. $\sqrt{901}$

13

© 2006 Walch Publishing

Solving Rational Number Equations

Some simple equations contain rational numbers in the form of fractions. These equations can be solved by finding ways of eliminating the fractions. For example, when an equation says $2x = 6$, then you can guess that the answer is 3. With an equation such as $\frac{3}{4}x = 3$, the answer is less obvious. One way to find the answer is to multiply the entire equation on both sides by the denominator of the fraction. This would be $(4)\left(\frac{3}{4}x\right) = (4)(3)$. The equation would then be $3x = 12$. It is easier to see now that the answer is 4.

14

Solve the following problems by eliminating the fractions.

1. $\frac{2}{3}x = 6$

2. $\frac{1}{2}x = 8$

3. $\frac{3}{4}x = 6$

4. $\frac{2}{5}x = 20$

© 2006 Walch Publishing

Rounding Decimals

When you buy gasoline, the price per gallon is usually given out to the thousandths place in dollars, which would be to the nearest $\frac{1}{10}$ of a cent. We don't pay the tenths of a cent, so we round to the nearest whole cent for the final cost. When rounding to a certain place, you look at the next smallest division. If that number is 4 or less, you do not round. If the number is 5 or more, you add one to the next division.

Round each number below to the indicated division.

1. 0.00045, to the thousandths place
2. 12.3466, to the hundredths place
3. 0.0388, to the hundredths place
4. 1.01010, to the hundredths place
5. 75.7774, to the tenths place
6. 5.14, to the tenths place
7. $16.884, to the nearest penny
8. $6.999, to the nearest penny
9. $8.20, to the nearest dime
10. $1599.3033, to the nearest penny

15

© 2006 Walch Publishing

Adding Decimals

Adding decimals allows us to collect many smaller parts of numbers and group them into larger totals that give us a more complete idea of the amount we're observing. For example, if you had 0.25 of one pie, 0.50 of another pie, and 0.125 of a third pie, would these parts all fit into a container meant to hold one pie? Add these three decimals together and find out.

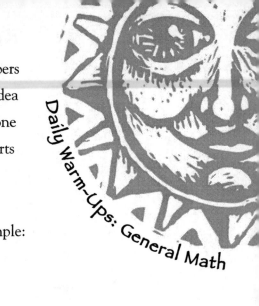

You can add decimals by lining up the decimal points. For example:

```
0.25 pie
0.50 pie
0.125 pie
0.875 pie    The total is less than one pie. All parts would fit into one
             container.
```

16

Add the following decimals. Be sure to line up the decimal points before you start.

1. 12.13, 2.005, 6.15
2. 0.001, 0.005, 0.022
3. 5.22, 5.23, 5.24
4. 6.5, 5.2, 2.1, 5.4, 8.7
5. 66.99998, 52.1114, 4.121214
6. 5.888, 4.555, 8.444

© 2006 Walch Publishing

Subtracting Decimals

Decimals can be thought of as parts or fractions of numbers. When adding decimals, you take many smaller parts and collect them into larger units. When subtracting using decimals, the smaller parts are taken from the larger unit to give you some idea of what is left. The most common place you see this in daily life is when you pay for something, generally with more money than the total of the purchase, and receive change. For example, if your restaurant bill is $14.36 and you pay with $15.00, you get back $0.64 in change. This subtraction of a decimal allows you to see how much you'll get back from the amount you had originally. Remember to line up the decimal points.

Solve the following problems.

1. 7.34 – 0.23

2. 5.41 – 0.12

3. 0.005 – 0.00478

4. 5 – 2.33

5. 10 – 0.01

6. 5.555 – 2.225

7. 65.001 – 3.003

8. 25.00 – 5.25

17

© 2006 Walch Publishing

Multiplying Decimals with Whole Numbers

You see many items in a store that have prices that are not whole numbers. For example, you might want to buy a candy bar that is $0.79 and then decide you would like to buy three more for some friends. How much will four candy bars cost in total? If you have $4.00 to spend, how much will you get back in change? Can you afford five candy bars with $4.00?

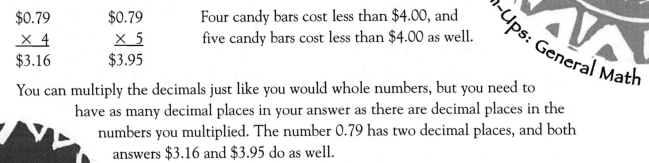

$0.79
× 4
―――
$3.16

$0.79
× 5
―――
$3.95

Four candy bars cost less than $4.00, and five candy bars cost less than $4.00 as well.

You can multiply the decimals just like you would whole numbers, but you need to have as many decimal places in your answer as there are decimal places in the numbers you multiplied. The number 0.79 has two decimal places, and both answers $3.16 and $3.95 do as well.

Solve the following problems.

1. 1.22 × 6

2. 0.004 × 3

3. 8 × 2.33

4. 3 × 0.344

5. 12 × 1.222

6. 0.515 × 9

18

© 2006 Walch Publishing

Multiplying Decimals with Decimals

Sometimes you buy items that have a decimal as part of their price. A pound of carrots might cost $2.25 per pound. What if you use a scale in the produce department and find that you bought 3.65 pounds of carrots? How do you determine the cost of the carrots at the checkout? You can multiply decimals using the same rules for multiplying whole numbers. The answer will contain the same number of decimal places as the total number of decimal places in the two numbers you multiplied together. For example:

$$\begin{array}{r} \$2.25 \\ \times\ 3.65 \\ \hline \$8.2125 \text{ (which rounds to \$8.21)} \end{array}$$

Note: $2.25 and $3.65 have two decimal places each, making a total of four decimal places just like you see in the answer.

Solve the following problems.

1. 1.01×1.02

2. 0.002×0.0222

3. 5.2×5.3

4. 12.001×8.002

5. 0.1×0.0001

6. 0.223×0.003

7. 12.1×3.95

8. 0.055×2.5

19

© 2006 Walch Publishing

Dividing Decimals with Whole Numbers

You may buy something at the store and find when you get home that you aren't sure what the unit cost was. (The unit cost is how much you paid per pound or per gallon, for example.) You can find out, however, if you know the total price you paid and how much of the item you bought. For example, if you bought 5 gallons of gasoline and the total was $13.80, you could determine the price per gallon as below.

$$5)\overline{13.80} = 2.76$$

Note: The decimal points line up in the answer and in the dividend.

Solve the following problems.

1. 21.6 ÷ 4

2. 0.002 ÷ 8

3. 15.25 ÷ 3

4. 2.555 ÷ 9

5. 0.0085 ÷ 5

6. 6.959 ÷ 6

20

© 2006 Walch Publishing

Dividing Decimals with Decimals

When dividing decimals by other decimals, it can be hard to keep track of where the decimal place should be in your answer. One of the ways to make the process easier is to turn your divisor (the number outside the division bracket) into a whole number. For example, if you wanted to solve the problem 6.555 ÷ 1.2, you could multiply both numbers by 10 and then the problem would be 65.55 ÷ 12. You can see below how the decimals change location. $1.2\overline{)6.555}$ would then become $12\overline{)65.55}$. The decimal in the divisor is moved to the right until it is a whole number, and then the decimal in the dividend is also moved to the right the same number of places. Only the divisor determines how far the decimals from both numbers are moved.

Solve the following problems. Round to five decimal places.

1. 21.644 ÷ 4.33

2. 0.002 ÷ 8.2

3. 15.25 ÷ 3.5

4. 2.555 ÷ 9.22

5. 0.0085 ÷ 0.0055

6. 5.959 ÷ 6.5

21

© 2006 Walch Publishing

Greatest Common Factor

When two or more whole numbers are multiplied together to form a product, those numbers are called **factors.** A number such as 100 has multiple factors, as does a number such as 75.

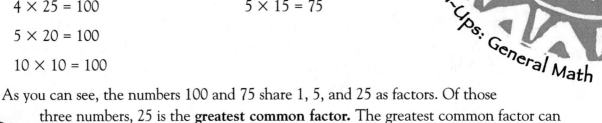

$1 \times 100 = 100$ $1 \times 75 = 75$

$2 \times 50 = 100$ $3 \times 25 = 75$

$4 \times 25 = 100$ $5 \times 15 = 75$

$5 \times 20 = 100$

$10 \times 10 = 100$

As you can see, the numbers 100 and 75 share 1, 5, and 25 as factors. Of those three numbers, 25 is the **greatest common factor.** The greatest common factor can come in handy when you are trying to simplify fractions.

Find the greatest common factor for each set of numbers below.

1. 12, 16, 20

2. 50, 75, 100

3. 2, 8, 64

4. 100, 200

5. 15, 45, 105

6. 225, 150, 300, 75

7. 55, 88, 121, 77

8. 62, 44, 38, 84

22

© 2006 Walch Publishing

Least Common Multiple

The **least common multiple** of two or more numbers can be found by listing all of the multiples of both numbers and finding the one of lowest value, excluding zero. For example, the multiples of 5 and 10 are listed below.

$1 \times 5 = 5$ $1 \times 10 = 10$

$2 \times 5 = 10$ $2 \times 10 = 20$

$3 \times 5 = 15$ $3 \times 10 = 30$

$4 \times 5 = 20$ $4 \times 10 = 40$

$5 \times 5 = 25$ $5 \times 10 = 50$

$6 \times 5 = 30$ $6 \times 10 = 60$

While 5 and 10 share 10, 20, and 30 as common multiples, the least common multiple is 10.

Find the least common multiple of each set of numbers below.

1. 8, 10

2. 2, 8

3. 7, 21

4. 24, 18

5. 12, 24, 36

6. 5, 8, 9

7. 2, 3, 5, 7

8. 45, 90, 180

23

© 2006 Walch Publishing

Simplifying Fractions

Fractions often appear in a form that is not as simple as possible. If 40 out of 100 people like a certain movie, then $\frac{40}{100}$ people like the movie. This fraction can be simplified by dividing out common factors until the only common factor is 1, or by dividing both numbers by the greatest common factor.

$\frac{40}{100} \div \frac{2}{2} = \frac{20}{50} \div \frac{2}{2} = \frac{10}{25} \div \frac{5}{5} = \frac{2}{5}$. This means that $\frac{40}{100} = \frac{2}{5}$, and 2 out of 5 people like the movie. If you had divided the numerator and the denominator by 20, the greatest common factor of the two numbers, you could have simplified the fraction in one step.

24

Simplify the following fractions.

1. $\frac{21}{81}$

2. $\frac{60}{100}$

3. $\frac{125}{500}$

4. $\frac{18}{20}$

5. $\frac{32}{64}$

6. $\frac{72}{360}$

© 2006 Walch Publishing

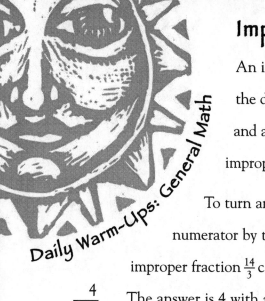

Improper Fractions to Mixed Numbers

An improper fraction is one in which the numerator is greater than the denominator. A **mixed number** contains both a whole number and a fraction. For example, $\frac{3}{2}$ is the same as $1\frac{1}{2}$, but $\frac{3}{2}$ is an improper fraction whereas $1\frac{1}{2}$ is a mixed number.

To turn an improper fraction into a mixed number, divide the numerator by the denominator and write the remainder as a fraction. So the improper fraction $\frac{14}{3}$ can be converted as below.

$$\begin{array}{r} 4 \\ 3\overline{)14} \\ \underline{12} \\ 2 \end{array}$$

The answer is 4 with a remainder of 2. The remainder is written over the divisor in fraction form, so the final answer is $4\frac{2}{3}$.

25

Write the following improper fractions as mixed numbers.

1. $\frac{28}{6}$

2. $\frac{16}{3}$

3. $\frac{21}{4}$

4. $\frac{50}{7}$

5. $\frac{83}{2}$

6. $\frac{44}{7}$

Daily Warm-Ups: General Math

© 2006 Walch Publishing

Mixed Numbers to Improper Fractions

Do you know how to change mixed numbers into improper fractions? The process is a little different from changing improper fractions to mixed numbers but can be done in fewer steps. For example, the mixed number $8\frac{1}{3}$ can be changed into an improper fraction by multiplying the denominator by the whole number, adding the numerator to the result, and taking the final answer and using it as the numerator over the original denominator. In other words, $3 \times 8 + 1 = 25$. The number 25 goes back over the 3 to make the improper fraction $\frac{25}{3}$.

26

Write the following mixed numbers as improper fractions.

1. $6\frac{1}{3}$

2. $8\frac{2}{5}$

3. $10\frac{2}{3}$

4. $7\frac{1}{4}$

5. $20\frac{6}{7}$

6. $1\frac{1}{3}$

7. $2\frac{5}{16}$

8. $11\frac{10}{11}$

© 2006 Walch Publishing

Ordering Fractions

Determining which of two fractions is larger can be simple if the fractions have like denominators, such as $\frac{1}{3}$ and $\frac{2}{3}$. It is easy to see that $\frac{2}{3}$ is the larger of the two. But how does $\frac{19}{27}$ compare with $\frac{2}{3}$? To answer this question, you need to write each fraction with a common denominator so you can compare numerators. For example, the least common denominator of $\frac{2}{3}$ and $\frac{19}{27}$ is 27. This means that if you multiply both the numerator and denominator of $\frac{2}{3}$ by 9, you get $\frac{18}{27}$. When you compare numerators, you see that $\frac{19}{27}$ is the larger of the two fractions.

Order the following fractions from smallest to largest by finding the least common denominator and then comparing numerators.

1. $\frac{2}{3}, \frac{3}{4}$

2. $\frac{3}{5}, \frac{3}{7}$

3. $\frac{2}{8}, \frac{3}{9}$

4. $\frac{1}{3}, \frac{2}{9}, \frac{5}{27}$

5. $\frac{1}{4}, \frac{1}{8}, \frac{17}{64}$

6. $\frac{2}{5}, \frac{9}{25}, \frac{32}{100}, \frac{17}{80}$

27

© 2006 Walch Publishing

Decimal to Fraction

Decimal numbers aren't always easy to picture. A decimal such as 0.40 is easy to multiply and divide, but it might not be easy to imagine 0.40 of a pie. As a fraction, the decimal 0.40 equals $\frac{2}{5}$, and some people can picture $\frac{2}{5}$ of a pie easier than 0.40 of a pie. To convert a decimal to a fraction, you can find the last decimal place value and express the decimal in fraction form to see if it can be simplified. For example, the last place value for the decimal 0.15 is the hundredths place, so you can express the decimal as $\frac{15}{100}$, which simplifies to $\frac{3}{20}$. If the decimal was 0.123, the last place would be the thousandths place, so you could express the decimal as $\frac{123}{1000}$.

Convert the following decimals to fractions and simplify where necessary.

1. 0.12
2. 0.15
3. 0.09
4. 0.58
5. 0.24

6. 0.125
7. 0.165
8. 0.10
9. 0.95
10. 0.64

28

© 2006 Walch Publishing

Fraction to Decimal

Some fractions are more difficult to use when you are carrying out a long line of mathematical operations. In some cases, it would be more convenient if the number was expressed as a decimal. The fraction $\frac{1}{2}$ is the same as the decimal 0.5, and you probably know a few others. To convert a fraction to a decimal, you need only divide the numerator by the denominator. For example:

$$\begin{array}{r} 0.5 \\ 2\overline{)1.0} \end{array}$$

Convert the following fractions to decimals, and round to the hundredths place.

1. $\frac{2}{3}$

2. $\frac{3}{4}$

3. $\frac{3}{5}$

4. $\frac{3}{7}$

5. $\frac{1}{3}$

6. $\frac{2}{9}$

7. $\frac{5}{27}$

8. $\frac{1}{4}$

9. $\frac{1}{8}$

10. $\frac{17}{64}$

11. $\frac{9}{25}$

12. $\frac{17}{80}$

29

© 2006 Walch Publishing

Adding and Subtracting Like Fractions

Like fractions are those fractions that have the same

denominator. For example, $\frac{1}{3}$, $\frac{2}{3}$, and $\frac{4}{3}$ are like fractions. To add

these fractions together, you add the numerators like whole

numbers and take the sum of the three numbers and place it above

the denominator, which is 3. So $\frac{1}{3} + \frac{2}{3} + \frac{4}{3} = \frac{7}{3}$. This is, of course, an

improper fraction and should be expressed as the mixed number $2\frac{1}{3}$.

Subtraction works in much the same way. For example: $\frac{5}{3} - \frac{7}{3} = -\frac{2}{3}$

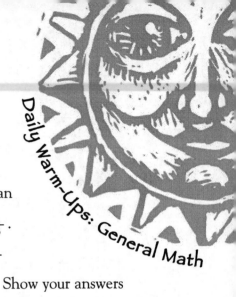

Complete the following addition and subtraction problems. Show your answers
as mixed numbers where necessary.

30

1. $\frac{2}{3} + \frac{5}{3}$

2. $\frac{3}{4} - \frac{1}{4}$

3. $\frac{3}{5} + \frac{2}{5}$

4. $\frac{3}{7} - \frac{8}{7}$

5. $\frac{1}{3} + \frac{5}{3}$

6. $\frac{5}{9} - \frac{1}{9}$

7. $\frac{5}{27} + \frac{50}{27}$

8. $\frac{1}{4} - \frac{70}{4}$

9. $\frac{1}{8} + \frac{6}{8}$

10. $\frac{17}{64} - \frac{9}{64}$

11. $\frac{9}{25} + \frac{13}{25}$

12. $\frac{17}{80} - \frac{35}{80}$

© 2006 Walch Publishing

Adding and Subtracting Unlike Fractions

Unlike fractions are those fractions that have different denominators, such as $\frac{2}{3}$, $\frac{3}{4}$, $\frac{3}{5}$, and $\frac{3}{7}$. When you add or subtract them, make sure they have common denominators by finding the least common denominator. Then add or subtract the numerators as you would whole numbers. The sum or difference of the numerators would be placed above the least common denominator. If the answer is an improper fraction, reduce it to a mixed number. For example, $\frac{2}{3}$ and $\frac{3}{4}$ would share 12 as the least common denominator. So the problem $\frac{2}{3} + \frac{3}{4}$ would become $\frac{8}{12} + \frac{9}{12} = \frac{17}{12}$, which would be the same as $1\frac{5}{12}$.

Solve the following addition and subtraction problems. Show your answers as mixed numbers where necessary.

1. $\frac{2}{3} + \frac{5}{4}$

2. $\frac{3}{4} - \frac{1}{5}$

3. $\frac{3}{5} + \frac{2}{6}$

4. $\frac{3}{7} - \frac{8}{9}$

5. $\frac{1}{4} - \frac{70}{9}$

6. $\frac{1}{8} + \frac{6}{13}$

7. $\frac{17}{64} - \frac{9}{40}$

8. $\frac{9}{25} + \frac{13}{8}$

31

© 2006 Walch Publishing

Multiplying Fractions

Multiplying fractions might come in handy in finding how big a piece of a piece might be. If you had half a cake and wanted to know how much a third of it would be, you would multiply $\frac{1}{2} \times \frac{1}{3}$. The numerators are multiplied, and their product stays on the top of the answer fraction. The denominators are also multiplied, and their product becomes the bottom of the answer fraction. For the problem above, that means the answer is $\frac{1}{6}$. Any answer that is an improper fraction should be converted to a mixed number.

Solve the following multiplication problems. Express the answers as mixed numbers where necessary.

32

1. $\frac{2}{3} \times \frac{5}{4}$

2. $\frac{3}{4} \times \frac{1}{5}$

3. $-\frac{3}{5} \times \frac{2}{6}$

4. $\frac{3}{7} \times \frac{8}{9}$

5. $\frac{1}{3} \times \frac{5}{4}$

6. $\frac{5}{9} \times -\frac{1}{10}$

7. $\frac{1}{4} \times \frac{70}{9}$

8. $-\frac{1}{8} \times \frac{6}{13}$

9. $\frac{17}{64} \times \frac{9}{45}$

10. $\frac{9}{25} \times \frac{13}{8}$

© 2006 Walch Publishing

Dividing Fractions

Reciprocals are two numbers that have a product of 1 when they are multiplied together. For example, 4 and $\frac{1}{4}$ are reciprocals. When dividing a fraction by a second fraction, you find that dividing a number by a fraction is the same as multiplying the number by the fraction's reciprocal. For example, $\frac{3}{4} \div \frac{1}{5} = \frac{15}{4}$ and $\frac{3}{4} \times \frac{5}{1} = \frac{15}{4}$. While the division problem may not be obvious, it is easier to see that in the second part you simply multiply the numerators to get the top number in the answer fraction and multiply the denominators to get the bottom number in the answer fraction. You can invert (flip over) the second fraction in the division problem to find its reciprocal and then can multiply the first fraction by that reciprocal.

Solve the following division problems. Express improper fractions as mixed numbers where necessary.

1. $\frac{2}{3} \div \frac{5}{4}$

2. $\frac{3}{4} \div \frac{1}{5}$

3. $\frac{3}{5} \div \frac{2}{6}$

4. $-\frac{3}{7} \div \frac{8}{9}$

5. $\frac{1}{3} \div \frac{5}{4}$

6. $\frac{5}{9} \div \frac{1}{10}$

7. $\frac{1}{8} \div \frac{6}{13}$

8. $-\frac{17}{64} \div \frac{9}{45}$

33

© 2006 Walch Publishing

Adding Mixed Numbers

Adding mixed numbers can be accomplished by following a short set of steps. First, add the fractions and simplify into a mixed number if necessary. Then add the whole numbers from each fraction as well as any whole number produced by the combination of the original fractions. For example, in the problem $1\frac{2}{5} + 3\frac{4}{5}$, $1 + 3 = 4$ and $\frac{2}{5} + \frac{4}{5} = \frac{6}{5}$, which is the same as $1\frac{1}{5}$. The 4 and $1\frac{1}{5}$ combine to equal $5\frac{1}{5}$. In this example, the fractions were like fractions, but in the event that the problem has unlike fractions, you only need to find the lowest common denominator to continue.

Solve the following problems.

1. $1\frac{1}{6} + 3\frac{1}{6}$

2. $2\frac{1}{7} + 4\frac{1}{7}$

3. $3\frac{3}{5} + 6\frac{3}{5}$

4. $1\frac{5}{8} + 1\frac{2}{8}$

5. $5\frac{1}{5} + 6\frac{3}{4}$

6. $15\frac{3}{7} + 15\frac{12}{14}$

34

© 2006 Walch Publishing

Subtracting Mixed Numbers

There are a number of ways to subtract mixed numbers. You can subtract the fractions first if they have like denominators and then subtract the whole numbers. You can convert the unlike fractions to have like denominators and then subtract fractions followed by whole numbers. A third way would be to make all mixed numbers into improper fractions, make like denominators, subtract, and then turn the answer into a mixed number if necessary.

Solve the following problems.

1. $1\frac{1}{6} - 3\frac{1}{6}$

2. $4\frac{1}{7} - 2\frac{1}{7}$

3. $3\frac{3}{5} - 6\frac{4}{5}$

4. $1\frac{5}{8} - 1\frac{2}{8}$

5. $6\frac{3}{5} - 5\frac{1}{5}$

6. $8\frac{1}{2} - 1\frac{1}{4}$

7. $15\frac{3}{7} - 15\frac{1}{7}$

8. $6\frac{1}{3} - 5\frac{1}{6}$

35

© 2006 Walch Publishing

Multiplying Mixed Numbers

Mixed numbers are best multiplied in their simplified form. For example, when multiplying $6\frac{3}{4}$ and $5\frac{1}{3}$, it would be easier to convert them to improper fractions before multiplying. The improper fractions would be $\frac{27}{4} \times \frac{16}{3} = \frac{432}{12} = 36$. By dividing out common factors at the beginning, you could have reduced the improper fractions and would have multiplied them as $\frac{9}{1} \times \frac{4}{1} = 36$. Solve the following problems.

1. $1\frac{1}{6} \times 3\frac{1}{2}$

2. $4\frac{1}{2} \times 2\frac{1}{3}$

3. $-3\frac{3}{5} \times 1\frac{1}{2}$

4. $1\frac{1}{8} \times 1\frac{1}{8}$

5. $2\frac{3}{4} \times 1\frac{1}{5}$

6. $2\frac{3}{7} \times -1\frac{1}{7}$

7. $1\frac{1}{2} \times 1\frac{2}{5}$

8. $2\frac{2}{3} \times 2\frac{1}{6}$

36

© 2006 Walch Publishing

Dividing Mixed Numbers

Dividing mixed numbers is similar to multiplying mixed numbers. The first step is to turn the mixed numbers into improper fractions and then find the reciprocal and multiply. The number of steps or the need for a calculator might be reduced if common factors are divided out after the reciprocal is set up, but before the multiplication step. For example, $6\frac{3}{2}$ divided by $1\frac{1}{2}$ as improper fractions would be $\frac{15}{2}$ and $\frac{3}{2}$. When multiplied by the reciprocal, this problem becomes $\frac{15}{2} \times \frac{2}{3}$. If you factor out the common factors, the problem becomes $\frac{5}{1} \times \frac{1}{1} = 5$.

Solve the following problems.

1. $1\frac{1}{6} \div 3\frac{1}{2}$

2. $4\frac{1}{2} \div 2\frac{1}{3}$

3. $-3\frac{3}{5} \div 1\frac{1}{2}$

4. $1\frac{1}{8} \div 1\frac{1}{8}$

5. $2\frac{3}{4} \div 1\frac{1}{5}$

6. $2\frac{3}{7} \div -1\frac{1}{7}$

7. $1\frac{1}{2} \div 1\frac{2}{5}$

8. $2\frac{2}{3} \div 2\frac{1}{6}$

37

© 2006 Walch Publishing

Adding Fractions and Mixed Numbers

When adding a fraction and a mixed number, one of the methods is to convert the mixed number into an improper fraction and add it to the other fraction. If the fractions are like fractions, they can be combined and the answer can be reduced to a mixed number. If the fractions are unlike fractions, then it will be necessary to find a common denominator before combining them and reducing the answer to a mixed number. For example:

$$\frac{3}{8} + 2\frac{1}{7} = \frac{3}{8} + \frac{15}{7} = \frac{21}{56} + \frac{120}{56} = \frac{131}{56} = 2\frac{19}{56}$$

This could also have been done by separating out the whole number, adding the fractions with like denominators, and then combining the answer back into a mixed number.

Solve the following problems.

1. $\frac{1}{6} + 3\frac{1}{6}$

2. $\frac{1}{5} + 1\frac{1}{7}$

3. $\frac{3}{5} + 2\frac{3}{5}$

4. $\frac{3}{4} + 1\frac{3}{8}$

5. $\frac{1}{5} + 2\frac{3}{4}$

6. $\frac{2}{3} + 1\frac{1}{7}$

7. $\frac{3}{7} + 1\frac{2}{5}$

8. $\frac{8}{9} + 2\frac{3}{5}$

38

© 2006 Walch Publishing

Subtracting Fractions and Mixed Numbers

When subtracting a fraction from a mixed number and vice versa, one of the methods is to convert the mixed number into an improper fraction before subtracting. If the fractions are like fractions, they can be subtracted, and the answer can be reduced to a mixed number. If the fractions are unlike fractions, then it will be necessary to find a common denominator before subtracting and reducing the answer to a mixed number. For example,

$$2\frac{1}{7} - \frac{3}{8} = \frac{15}{7} - \frac{3}{8} = \frac{120}{56} - \frac{21}{56} = \frac{99}{56} = 1\frac{43}{56}$$

This could also have been done by separating out the whole number, subtracting the fractions with like denominators, and then combining the answer back into a mixed number.

Solve the following problems.

1. $3\frac{1}{6} - \frac{1}{6}$

2. $1\frac{1}{7} - \frac{1}{5}$

3. $2\frac{3}{5} - \frac{3}{5}$

4. $1\frac{3}{8} - \frac{3}{4}$

5. $2\frac{3}{4} - \frac{1}{5}$

6. $1\frac{1}{7} - \frac{2}{3}$

7. $1\frac{2}{5} - \frac{3}{7}$

8. $2\frac{3}{5} - \frac{8}{9}$

39

© 2006 Walch Publishing

Multiplying Fractions and Mixed Numbers

Mixed numbers and fractions are best multiplied in their simplified form. For example, when multiplying $\frac{3}{4}$ and $4\frac{1}{2}$, it would be easier to convert the mixed number to an improper fraction before multiplying. The improper fraction would be $\frac{9}{2}$, and the problem could be solved by $\frac{3}{4} \times \frac{9}{2} = \frac{27}{8} = 3\frac{3}{8}$. In some cases when the fractions allow, the problem could be simplified if you were able to divide out some common factors before multiplying.

Solve the following problems.

1. $1\frac{1}{6} \times \frac{2}{5}$

2. $\frac{1}{7} \times 2\frac{1}{3}$

3. $3\frac{1}{4} \times \frac{4}{5}$

4. $\frac{5}{8} \times 1\frac{2}{3}$

5. $1\frac{1}{4} \times \frac{1}{5}$

6. $\frac{2}{3} \times 1\frac{1}{7}$

7. $1\frac{3}{7} \times \frac{2}{5}$

8. $\frac{8}{9} \times 5\frac{1}{3}$

Daily Warm-Ups: General Math

40

© 2006 Walch Publishing

Dividing Fractions and Mixed Numbers

Dividing a mixed number by a fraction and vice versa is similar to multiplying. The first step is to turn the mixed number into an improper fraction. Then find the appropriate reciprocal and multiply. The number of steps or the need for a calculator might be reduced if common factors are divided out after the reciprocal is set up, but before the multiplication step. For example, $6\frac{3}{2}$ divided by $\frac{5}{6}$ would become $\frac{15}{2}$ divided by $\frac{5}{6}$. When you multiply by the reciprocal, this problem becomes $\frac{15}{2} \times \frac{6}{5}$. If you factor out the common factors, the problem becomes $\frac{3}{1} \times \frac{3}{1} = 6$.

Solve the following problems.

1. $1\frac{1}{6} \div \frac{2}{5}$

2. $\frac{1}{7} \div 2\frac{1}{3}$

3. $3\frac{1}{4} \div \frac{4}{5}$

4. $\frac{5}{8} \div 1\frac{2}{3}$

5. $1\frac{1}{4} \div \frac{1}{5}$

6. $\frac{2}{3} \div 1\frac{1}{7}$

7. $1\frac{3}{7} \div \frac{2}{5}$

8. $\frac{8}{9} \div 5\frac{1}{3}$

41

© 2006 Walch Publishing

Proportions

A **proportion** is an equation that says that two ratios are the same. For example, $\frac{1}{2} = \frac{8}{16}$ is a proportion because the equation is true. Another way to check to see if two fractions form a proportion is to see if their cross products are equal. This means that the denominator from the first fraction times the numerator from the second fraction will have a product equal to the numerator from the first fraction times the denominator from the second fraction. In the example above, you can see that $2 \times 8 = 1 \times 16$.

Circle the proportions that are correct.

1. $\frac{1}{2} = \frac{3}{16}$ 2. $\frac{1}{3} = \frac{2}{6}$ 3. $\frac{1}{4} = \frac{2}{8}$ 4. $\frac{1}{5} = \frac{6}{25}$

42

Find the missing number that makes each proportion correct.

5. $\frac{2}{7} = \frac{x}{21}$ 6. $\frac{3}{8} = \frac{21}{x}$ 7. $\frac{4}{9} = \frac{x}{27}$ 8. $\frac{5}{12} = \frac{125}{x}$

© 2006 Walch Publishing

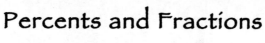

Percents and Fractions

25% means 25 parts out of 100. This is the same as saying $\frac{25}{100}$, which reduces to $\frac{1}{4}$. Fractions are another way of writing percents, and percents are another way of writing fractions. Any percent can be written as a fraction. Just take the percent and put it into a fraction with 100 as the denominator and then simplify. Any fraction can be written as a percent by setting up a proportion with the fraction and a second fraction with a denominator of 100. Solving for the missing numerator above the 100 will give you the number that will become the percent. For example, setting $\frac{3}{4}$ equal to $\frac{x}{100}$ you find that x is 75, so the fraction $\frac{3}{4}$ is the same as 75%.

Change each percent to a fraction.

1. 12.5%

2. 28%

3. 16%

4. 21%

5. 80%

Change each fraction to a percent.

6. $\frac{1}{7}$

7. $\frac{2}{25}$

8. $\frac{7}{20}$

9. $\frac{9}{10}$

10. $\frac{2}{5}$

43

© 2006 Walch Publishing

Percents and Decimals

Percents and decimals are numbers that have the same value, but different meanings. A decimal can be changed into a percent simply by multiplying the decimal by 100 and adding a percent sign. A percent can be changed into a decimal by the opposite action of dropping the percent sign and dividing the number by 100.

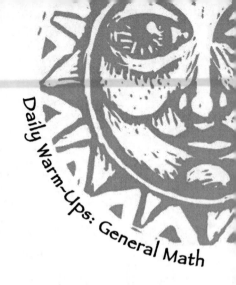

Convert each decimal to a percent.

1. 0.25

2. 0.75

3. 0.66

4. 1.15

5. 0.89

Convert each percent to a decimal.

6. 12%

7. 61%

8. 12.50%

9. 15.2%

10. 99.99%

44

© 2006 Walch Publishing

Unusually Large Percents

Sometimes you will see percents that are greater than 100%. You might wonder how you can have more than all of something, but it's really a way of comparing two or more things. For example, if you get up every morning and run 4 miles, then one day you run 5 miles, what percent of your daily run did you do? In this case, you ran $\frac{5}{4}$ of your normal run. As a decimal, that number would be 1.2. To convert the decimal to a percent, you multiply by 100 and add a percent sign to get 120%. A percent that is greater than 100% is not wrong; it usually just indicates that a comparison is being made.

Write each item as the following:

decimal	fraction	percent
1. 135%	5. 1.5	9. $\frac{10}{3}$
2. 1500%	6. 2.8	10. $\frac{12}{11}$
3. $\frac{24}{5}$	7. 175%	11. 3.5
4. $\frac{7}{3}$	8. 125%	12. 4.66

45

© 2006 Walch Publishing

Unusually Small Percents

Sometimes percents will be less than 1%. How can you have less than 1 of something? It is a way of pointing out a very small amount compared to the total amount. For example, if you walk to school exactly 1 mile (5280 feet) from your house, and one day you walk 10 feet before a neighbor gives you a ride to school, what percent of your normal trip did you walk? In this case, you walked $\frac{10}{5280}$ of your normal walk. As a decimal, that number would be 0.00189. To convert the decimal to a percent, multiply by 100 and add a percent sign to get 0.189%. A percent that is less than 1% isn't wrong. It usually indicates the percent is very small compared to the total.

Write each percent as a decimal.

1. 0.05%

2. 0.25%

3. 0.033%

4. 0.99%

5. 0.12%

Write each decimal as a percent.

6. 0.002

7. 0.0015

8. 0.0033

9. 0.0068

10. 0.0009

© 2006 Walch Publishing

Percent of a Number

Human beings like to be able to classify things into categories that make them easier to understand. For example, you might wonder how many people own dogs in your town and take a poll to find out. You discover that 24% of households own a dog. If your town has 12,623 homes, how many households is that? You can find out by converting the percent to a decimal and then multiplying by the number of homes. The problem would be 0.24 × 12,623 = 3029.52, or rounded to 3030 households. This process can be used to find the percent of any number.

Solve the following problems.

1. 12% of 365

2. 75% of 60

3. 33% of 900

4. 45% of 11

5. 99% of 662

6. 8% of 49

7. 11% of 121

8. 3.4% of 78

9. 56% of 95.45

10. 12.5% of 80

11. 12.5% of 88

12. 16.6% of 1678.87

47

© 2006 Walch Publishing

The Percent Proportion

When you make a proportion, you make a connection between two fractions of equal value (for example, $\frac{1}{2} = \frac{2}{4}$). In a percent proportion, the first fraction is $\frac{\text{part of total}}{\text{total}}$ and the second fraction is always $\frac{\text{percent}}{100}$. The "part of total" is some small part of a larger number, and the "total" is the whole amount of the number. You can fill in numbers for these values and can then use a cross product to solve. For example, $\frac{7}{12} = \frac{\text{percent}}{100}$ so $(7)(100) = (12)(\text{percent})$. If you divide both sides by 12, you find that the percent is 58.3%. You can also do this for a missing part of total or total. For example, $\frac{\text{part of total}}{45} = \frac{33}{100}$. So with a cross product you find that $(33)(45) = (100)(\text{part of total})$. If you divide both sides by 100, you find the part of total is 14.85.

Use the percent proportion to solve the following problems.

1. $\frac{7}{15} = \frac{\text{percent}}{100}$

2. $\frac{7}{\text{total}} = \frac{35}{100}$

3. $\frac{\text{part of total}}{65} = \frac{14}{100}$

4. $\frac{85}{785} = \frac{\text{percent}}{100}$

5. $\frac{25.8}{\text{total}} = \frac{155}{100}$

6. $\frac{\text{part of total}}{901} = \frac{12.89}{100}$

48

© 2006 Walch Publishing

The Percent Equation

The percent equation is a way of performing a percent proportion, but with fewer steps. Because the second fraction is always $\frac{\text{percent}}{100}$, you can save a step by expressing the percent as a decimal from the beginning of a problem. The cross product becomes (part of total) = (percent)(total). This form is called the percent equation. For example, if you are trying to find 15% of 65, 15% is the percent and 65 is the total. So the percent equation would read: (part of total) = (0.15)(65) = 9.75. No matter what a question asks about percent, total, and part of total, you can use this formula if you know any two of the parts.

Solve the following problems using the percent equation.

1. What is 32% of 68?

2. 45 is what percent of 93?

3. 6 is 25% of what number?

4. What is 12.5% of 155?

5. 90 is what percent of 125?

6. 98 is 16% of what number?

7. What is 0.04% of 850?

8. 62 is what percent of 180?

9. 3.44 is 85% of what number?

10. What is 114% of 70?

11. 52 is what percent of 30?

12. 40 is 251% of what number?

49

© 2006 Walch Publishing

Percent Change

If you were 20 inches long when you were born and are now 60 inches, you are 40 inches taller. But what percent increase is that in your height? If 20 was your original height, that used to represent 100% of your height, then 40 would be twice as much, or 200%. You have grown 200% from your original height. You can write this percent increase as the following formula:

percent change $= \frac{\text{amount of change}}{\text{original amount}}$. For your height, the formula would be the following: percent change $= \frac{40}{20} = 2$, which equals 200%.

Solve the following problems using the percent change formula.

1. You have 20 pencils and give away 7. What is the percent change?

2. You have 30 marbles and a friend gives you 26 more. What is the percent change?

3. You were 55 inches tall last year, and now you're 68 inches tall. What is the percent change in your height?

4. A tree near your school was planted when it was 4 feet tall, and now it is 35 feet tall. What is the percent change in the tree's height?

© 2006 Walch Publishing

Sales Tax

Many of the percent, decimal, and fraction problems you work on in math revolve around money. One of the things that can be confusing when finding the cost of an item in a store is sales tax. Sales tax is usually a small amount of the total cost of an item that goes to the government to help pay for programs. For example, if you buy an item that costs $1 in a state with a 5% sales tax, you have to pay an extra 5 cents. Sales tax can be found in one of two ways. You can multiply the cost of the item by the percent sales tax as a decimal to find just the tax. You can also multiply the cost of the item by 1 plus the percent sales tax as a decimal to find the total cost with tax.

For example, 5% sales tax on an item that costs $4.60 would amount to (0.05)($4.60) = $0.23. To find the total cost, you could use the second method described above: (1.05)($4.60) = $4.83.

Find the sales tax charged for each item.

1. 8% tax on a $12.56 item

2. 5% tax on a $50 item

3. 6.5% tax on a $3.25 item

4. 17% tax on a $8.99 item

Find the cost of each item.

5. 3% tax on a $158.75 item

6. 4.5% tax on a $22.60 item

7. 7.5% tax on a $45.99 item

8. 2.25% tax on a $32.19 item

51

© 2006 Walch Publishing

Sales and Discounts

If an item is on sale or offered at a discount, the amount that the price is reduced is usually given as a percent. If a camera that costs $100 is offered at 25% off, you probably know that the final price is $75. But how do you figure out the final price using other less common percents? Probably the simplest way is to take the sales percent and subtract it from 100%. The difference is then converted to a decimal and multiplied by the cost of the item. For example, if an MP3 player that costs $150 is on sale for 33% off, what is the sale price? 100% – 33% = 67% = 0.67, which you multiply by the original price. (0.67)($150) = $100.50, which is the sale price. To find the discount, you subtract the sales price from the original price: $150 – $100.50 = $49.50.

Find the sale price of each item below.

1. 10% off $120

2. 20% off $45.50

3. 25% off $65.00

4. 33% off $3000

Find the discount of each item below.

5. 40% off $2500

6. 15% off $29.99

7. 18% off $39.99

8. 60% off $16.95

52

© 2006 Walch Publishing

Simple Interest

When money is placed into a savings account, the bank pays a certain amount of interest. **Interest** is money paid to you because the bank is using your money to do other jobs. Sometimes you pay interest to someone when you are using his or her money, such as if you borrow money to pay for a car. You might also pay interest on a credit card or to a store where you have a credit account. You can calculate the amount of interest by multiplying the principal (the amount borrowed) by the interest rate, which is given as a percent. For example, if you have $450 in a savings account that pays 2.5% interest per year, you would multiply ($450)(0.025) = $11.25.

Calculate the interest for each problem below.

1. 3% interest earned on a $500 savings account

2. 5% interest earned on a $1500 savings bond

3. 12% interest owed on a $20,000 car loan

4. 22% interest owed on a $5500 credit card bill

5. 2.5% interest earned on $5,800,000 lottery winnings

6. 10% interest owed on a $255.99 store credit account

53

© 2006 Walch Publishing

Compound Interest

If you put $100 in an account that earns 5% interest per year, at the end of the year you would have the original $100 plus $5 earned as interest. If you left the money for another year, you would earn the 5% interest on the entire amount of $105. The interest earned over four years is shown below.

Year 1: $100 × 1.05 = $105
Year 2: $105 × 1.05 = $110.25
Year 3: $110.25 × 1.05 = $115.76
Year 4: $115.76 × 1.05 = $121.55

The formula for compound interest, compounded yearly, is Total = (Original Principal)(1 + interest rate)^n, or $T = P(1 + r)^n$. The n in the formula represents the number of years the interest is growing. *Note:* The interest rate, which is given as a percent, needs to be converted to a decimal before it is added to the 1 in the parentheses.

54

Calculate the totals for the following problems.

1. $100 invested at 7% for 10 years

2. $10,000 invested at 3% for 5 years

3. $50,000 invested at 2% for 40 years

4. $300 invested at 3.5% for 12 years

5. $250 invested at 40% for 5 years

© 2006 Walch Publishing

The Metric System

The **metric system** was developed in the 1790s by a group of scientists. They wanted to be sure a measurement made in one place on Earth could be reproduced on any other part of Earth. The modern version of the system is called the SI system from the French phrase *Le Système International d'Unité.* However, many people still call it the metric system. The system uses meters to measure length, liters to measure volume, and grams to measure mass. The system also uses divisions of ten instead of fractions. For example, a decimeter is $\frac{1}{10}$ of a meter, and a decameter is 10 meters. The different divisions are indicated with different prefixes such as deci-, deca-, milli-, micro-, mega-, giga-, and many others.

Why do you think scientists wanted a measurement system that was the same for all people from all countries?

55

© 2006 Walch Publishing

Metric Length

The basic unit of length in the metric system is the meter. Different prefixes are put in front of the word *meter* to indicate different-sized metric measurements. For example, a millimeter is about the thickness of a dime. A centimeter is about the height of a stack of 5 nickels. A meter is a little more than 3 feet. A kilometer is about 3281 feet or about 0.62 miles.

Choose the best unit, a millimeter (mm), a centimeter (cm), a meter (m), or a kilometer (km), to measure each of the following things.

1. your height

2. the thickness of a telephone book

3. the distance across a gymnasium

4. the distance from New York City to Los Angeles

5. the height of a car

6. the width of a pencil

7. the thickness of a piece of spaghetti

8. the length of a football field

56

© 2006 Walch Publishing

Metric Volume

The basic unit of volume in the metric system is the liter. You have probably seen a 1-liter bottle or a 2-liter bottle of soda or water at the store or in your home. The size of a liter is defined as the amount of space occupied by 1 cubic decimeter. Remember that a decimeter is 10 centimeters, so that would be a box that is 10 centimeters tall, 10 centimeters wide, and 10 centimeters deep. A milliliter is $\frac{1}{1000}$ of a liter and is about the size of 10 to 15 water droplets. You often see milliliters given as a unit of measure on medicines. A millimeter is also the same size as a cubic centimeter, or cc, which you might hear mentioned on medical shows or in a discussion about small engines.

Choose the best unit, liters (l) or milliliters (ml), to measure each of the following.

1. a swimming pool filled with water

2. a syringe

3. a tablespoon

4. a soda can

5. a 5-gallon bucket

6. a handful of water

7. the amount of air in a tire

8. the amount of air in a balloon

9. the amount of air in your ear

10. the amount of coffee in a mug

57

© 2006 Walch Publishing

Metric Mass

The base unit of mass in the metric system is the kilogram. A kilogram is about 2.2 pounds, which is also the mass of 1 liter of water. Some other common units are the gram, which is about the mass of a paper clip. A nickel has a mass of almost exactly 5 grams. A milligram is very small and would be about the mass of a piece of sand or a grain of salt.

Choose the best unit, kilograms (kg), grams (g), or milligrams (mg), to measure each of the following.

1. the mass of a person

2. the mass of a candy bar

3. the mass of a car

4. the mass of an ant

5. the mass of a desk

6. the mass of a feather

7. the mass of a dime

8. the mass of a textbook

58

© 2006 Walch Publishing

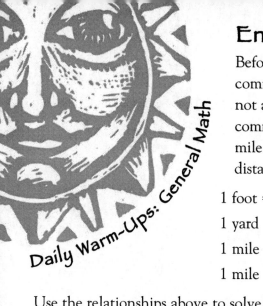

English Length

Before the metric system, the English system of units was more common. The English system is still used in some countries, but not as widely as the metric system. In the English system, the common units of length are the inch, the foot, the yard, and the mile. Below are some of the relationships between the units of distance.

1 foot = 12 inches

1 yard = 3 feet

1 mile = 5280 feet

1 mile = 1760 yards

Use the relationships above to solve the following problems.

1. Change 48 inches to feet.

2. Change 36 feet to yards.

3. Change 12 miles to yards.

4. Change 6 miles to feet.

5. Change 1 mile to inches.

6. Change 750 inches to yards.

7. Change 651 yards to feet.

8. Change 7.5 yards to inches.

9. Change 15,000 feet to miles.

10. Change 25,680 yards to miles.

11. Change 349 feet to inches.

12. Change 89,442 inches to miles.

59

© 2006 Walch Publishing

English Volume

The English system of units has a large number of measures of volume. The system has many unusual terms such as barrels, bushels, hogsheads, and gills. In the English system, the common units of volume are the ounce, the cup, the pint, the quart, and the gallon. Below are some of the relationships between the units of volume.

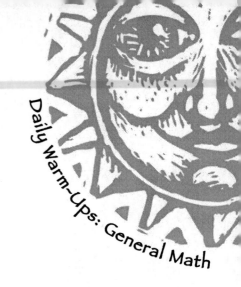
1 cup = 8 ounces

1 pint = 2 cups

1 quart = 2 pints

1 gallon = 4 quarts

60

Use the relationships above to solve the following problems.

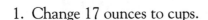

1. Change 17 ounces to cups.

2. Change 55 ounces to pints.

3. Change 2 gallons to quarts.

4. Change 16 gallons to pints.

5. Change 12 pints to cups.

6. Change 6 gallons to cups.

7. Change 6.5 quarts to ounces.

8. Change 12 gallons to ounces.

9. Change 18 quarts to cups.

10. Change 80 pints to quarts.

© 2006 Walch Publishing

English Weight

The English system of units is still used in some countries and has a fairly small number of weight units. In the English system, the common units of weight are the ounce, the pound, and the ton. People who work with large weights also use two kinds of ton, the short ton and the long ton. Below are some of the relationships between the units of weight.

1 pound = 16 ounces
1 short ton = 2000 pounds
1 long ton = 2240 pounds

Use the relationships above to solve the following problems.

1. Change 12 pounds to ounces.

2. Change 3.5 short tons to pounds.

3. Change 1.6 short tons to ounces.

4. Change 20 long tons to short tons.

5. Change 35 short tons to long tons.

6. Change 3.4 long tons to pounds.

7. Change 0.005 short tons to ounces.

8. Change 78,500 ounces to pounds.

9. Change 950,000 ounces to long tons.

10. Change 3.5 long tons to pounds.

61

© 2006 Walch Publishing

Time

Both the metric system and the English system use the same basic units of time. Those units are the second, the minute, the hour, the day, and the year. There are other units such as the month, the week, and the leap year. Some terms are better than others for measuring or describing time. For example, you usually give your age in years, but very seldom in minutes. Below are some of the relationships between the units of time.

1 minute = 60 seconds

1 hour = 60 minutes

1 day = 24 hours

1 year = 365 days

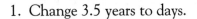

Use the relationships above to solve the following problems.

62

1. Change 3.5 years to days.

2. Change 2.66 years to hours.

3. Change 12 hours to minutes.

4. Change 72 minutes to seconds.

5. Change 87,500 seconds to days.

6. Change 4588 days to years.

7. Change 4.4 years to minutes.

8. Change 700 hours to seconds.

9. Change 1 year to seconds.

10. Change 64 days to minutes.

© 2006 Walch Publishing

Converting Metric and English Units

The metric system and the English system are still used in many places. Miles and kilometers appear side by side on car speedometers. Some gas stations sell gasoline by the gallon and by the liter. Soda cans are marked in both ounces and milliliters. Food in cans is often labeled in both ounces and grams. Below are some common metric and English terms and how they are related.

1 kilogram = 2.2 pounds	1 mile = 1.6 kilometers
1 gallon = 3.78 liters	1 inch = 2.54 centimeters
1 liquid ounce = 29.57 milliliters	

Use the relationships above to solve the following problems.

1. Change 2.4 kilograms to pounds.

2. Change 16 pounds to kilograms.

3. Change 6.6 gallons to liters.

4. Change 55.8 liters to gallons.

5. Change 75 liquid ounces to milliliters.

6. Change 8700 milliliters to fluid ounces.

7. Change 25 miles to kilometers.

8. Change 350 kilometers to miles.

9. Change 12 inches to centimeters.

10. Change 733 centimeters to inches.

63

© 2006 Walch Publishing

Scientific Notation

Scientific notation takes a number and expresses it as a number from 1 to 9 multiplied by some power of 10. Any number can be expressed in scientific notation. Generally it is used for very large or very small numbers. For example, 100 can also be written as 10^2. This means that 200 can be written as 2×100 or 2×10^2. 3500 can be written as 3.5×1000, or 3.5×10^3.

Write each number below in scientific notation.

1. 75,000
2. 1,300,000
3. 5500

4. 8,000,000,000,000
5. 0.0000000095

Write each number below in standard form.

6. 7.3×10^4
7. 2×10^6
8. 3.66×10^8

9. 1.2×10^2
10. 9.99×10^{-7}

64

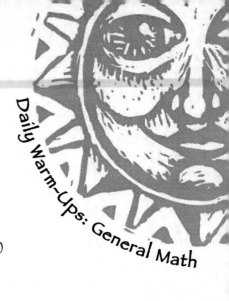

© 2006 Walch Publishing

Significant Digits

If you were to weigh a small rock on a scale that could measure the mass of the rock to the nearest 0.01 g, then the mass of the rock would be, for example, 5.22 ± 0.01 g. The last digit is really just the best estimate of what the last digit should be. Perhaps it was rounded or perhaps not, but since there is no way to be certain, the last digit is called uncertain. The first two digits are not estimated. These digits are called **significant figures.** There are four requirements for significant figures: 1. All nonzero numbers are significant; 2. Zeros between nonzeros are significant; 3. Placeholding zeros at the beginning and end of a number are not significant; 4. Zeros at the end of a number after the decimal are significant.

How many significant figures are in each of the following?

1. 5.01

2. 80.5

3. 70

4. 3002

5. 0.0016

6. 0.7070400800

7. 12.45

8. 12.0450

65

© 2006 Walch Publishing

Precision

The ruler below is measuring a line that is at least 8 centimeters long, but not quite 9 centimeters long. Is it 8.5 centimeters? Is it 8.6 or 8.7? **Precision** is a measurement of how exactly a measurement was made. The precision of this ruler is 1 centimeter. Any smaller measurements must be estimated. It would not be wise to estimate down to the hundredths place because even the numbers in the tenths are guesses, so a number such as 8.66 would be a fairly wild guess.

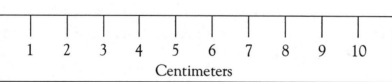

66

Give the unit of precision for each of the following items.

1. car odometer (measures distance)

2. car speedometer

3. bathroom scales

4. gasoline pump

5. digital thermometer for measuring your temperature

6. meterstick

7. 1-foot-long ruler

8. supermarket scales

© 2006 Walch Publishing

Frequency Tables

A **frequency table** is a way of organizing many numbers that have been collected or written down in no particular order. For example, if you asked each person who walked into your local supermarket in which month he or she was born, you would not get all the answers in order. You would make a frequency table and collect the numbers as they arrived. The first part of the table (January to May only) might look something like the table below.

Month	Number collected	Frequency
January	ЖЖ II	7
February	III	3
March	ЖЖ II	7
April	ЖЖ III	8
May	ЖЖ III	8

Make a frequency table for all the students in your class that summarizes everyone's height in inches.

67

© 2006 Walch Publishing

Bar Graphs

A **bar graph** is best used when collecting a lot of information that has been counted. For example, you might count the number of pieces of mail that arrive at your house each day.

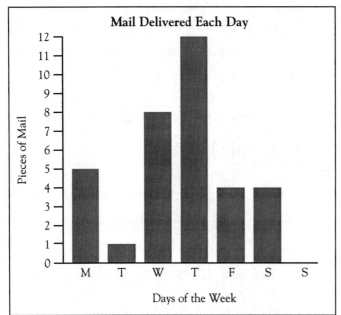

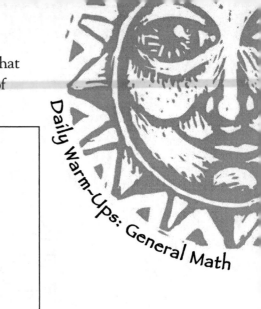

Find out what month each of your classmates was born in and make a bar graph displaying the data.

© 2006 Walch Publishing

Circle Graphs

A **circle graph** is best used to show how some large collection of data is cut into pieces. The pieces are often labeled in percents and look like slices of pie, which is why the circle graph is sometimes called a pie graph or a pie chart.

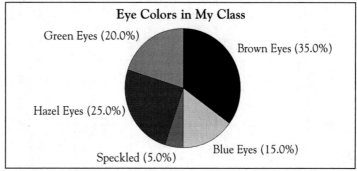

Eye Colors in My Class

Green Eyes (20.0%)

Brown Eyes (35.0%)

Hazel Eyes (25.0%)

Blue Eyes (15.0%)

Speckled (5.0%)

Answer the following questions about the chart above.

1. What does this circle graph show?

2. What eye color is most common? Least common?

3. What percent is the most common eye color? The least common?

4. Make a circle graph of eye colors in your class and decide which colors should and should not be on the list.

69

© 2006 Walch Publishing

Line Graphs

Line graphs are best used to show a change over time. They allow us to collect data that can be used to predict how things may change in the future.

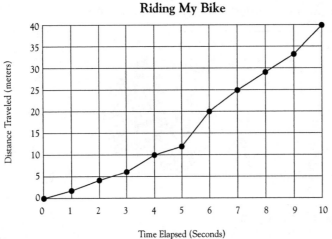

Riding My Bike

Answer the following questions about the graph above.

1. How far did the biker travel after 4 seconds? After 6.5 seconds?

2. How long did it take the biker to travel 25 meters? 40 meters?

3. Based on the available information, how long should it take the biker to travel 100 meters? Is this an accurate number or just a guess?

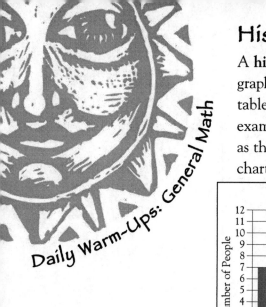

Histograms

A **histogram** is a way of combining a frequency table and a bar graph. The data is either collected or organized on a frequency table and then the results are transferred to a bar graph. For example, if you collected the birth months of a number of people as they passed you in a mall, you could order them on a frequency chart and then graph the results in a histogram, as below.

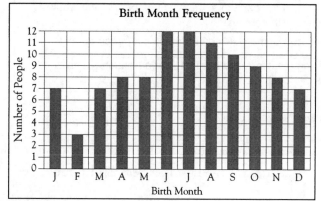

Answer the following questions about the histogram above.

1. When were the fewest people born? The most people born?

2. What part of the year does the birthrate seem to increase? Decrease?

3. What percent of the people were born in February? In June?

© 2006 Walch Publishing

Which Graph Is Best?

In the chart below are the five basic ways of organizing data. Here is a summary of their strengths.

Graph type	Best used for
frequency chart	organizing data that falls into categories and is being counted
bar graph	organizing data that has already been counted
circle graph	showing how the parts of something compare to the whole; often used with percents
line graph	showing how some collected data has changed over time
histogram	combining the information collected in a frequency chart and organizing it into a bar graph

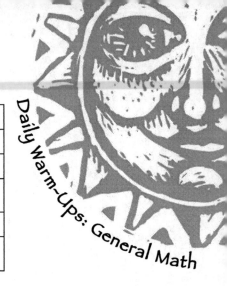

Choose which chart above is best for showing the data below.

1. You count the number of trucks, cars, motorcycles, bicycles, and pedestrians that pass through an intersection in one hour.

2. You want to show how you've grown over the last ten years.

3. You want to show the percent makeup of cake ingredients.

4. You show how many baseball teams use a color in their uniform.

72

© 2006 Walch Publishing

Matrices

A **matrix** is a way of arranging data into a rectangle made up of rows and columns. Putting data into a matrix is a way of organizing the material so that it can be combined with other matrices. For two matrices to be added or subtracted, they must have the same dimensions. This means that they must have the same number of rows and the same number of columns. If this is true, then the matrices can be added or subtracted by adding or subtracting numbers that are in the same position in each matrix. For example:

$$\begin{bmatrix} 2 & 8 & 4 \\ 5 & 8 & 0 \\ 6 & 3 & 9 \end{bmatrix} + \begin{bmatrix} 7 & 2 & 3 \\ 2 & 4 & 8 \\ 3 & 6 & 12 \end{bmatrix} = \begin{bmatrix} 2+7 & 8+2 & 4+3 \\ 5+2 & 8+4 & 0+8 \\ 6+3 & 3+6 & 9+12 \end{bmatrix} \text{ which simplifies to } \begin{bmatrix} 9 & 10 & 7 \\ 7 & 12 & 8 \\ 9 & 9 & 21 \end{bmatrix}$$

Solve the following problems.

1. $\begin{bmatrix} 1 & 3 \\ 6 & 8 \end{bmatrix} + \begin{bmatrix} 2 & 4 \\ 8 & 14 \end{bmatrix}$

3. $\begin{bmatrix} 1 & 3 \\ 6 & 8 \end{bmatrix} + \begin{bmatrix} 7 & 11 & 3 \\ 4 & 9 & 1 \\ 2 & 5 & 12 \end{bmatrix}$

2. $\begin{bmatrix} 7 & 11 & 3 \\ 4 & 9 & 1 \\ 2 & 5 & 12 \end{bmatrix} + \begin{bmatrix} 5 & 6 & 4 \\ 5 & 11 & 4 \\ 5 & 18 & 23 \end{bmatrix}$

4. $\begin{bmatrix} 5 & 5 & 3 \\ 6 & 11 & 8 \\ 4 & 4 & 5 \end{bmatrix} - \begin{bmatrix} 2 & 2 & 1 \\ 12 & 25 & 14 \\ 2 & 1 & 1 \end{bmatrix}$

73

© 2006 Walch Publishing

Perimeter

Perimeter is the distance around a closed shape. For example, a square might be 4 centimeters long on each side, which means that the total length of all four sides is 16 centimeters. 16 centimeters is the perimeter of that square. The perimeter of any shape can be found if you know the length of all the sides. If you know the lengths, you can simply add them together.

Find the perimeter of each shape below.

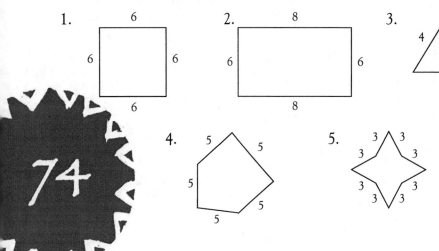

1.
```
    6
6       6
    6
```

2.
```
    8
6       6
    8
```

3.
```
  /\
4/  \4
/____\
   4
```

4.
```
  5
5    5
5      5
   5
```

5.
```
  3 /\ 3
3      3
3      3
  3 \/ 3
```

6.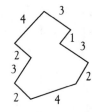
```
4    3
   1 3
2
3
2
  4    2
```

74

Area of a Rectangle

A **rectangle** is a figure that has four sides. A rectangle must also have four right angles (90°) at the corners. Sides of a rectangle that are opposite each other must also be the same length. In most rectangles, the longer direction is called the length and the shorter direction is called the width. The area of a rectangle can be found by multiplying the length times the width. The formula for the area of a rectangle is area equals length time width, or $A = l \cdot w$.

Calculate the area of each rectangle below.

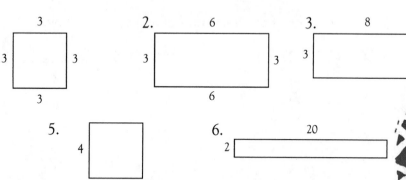

1.
3
3   3
3

2.
6
3   3
6

3.
8
3

4.
6
2

5.
4

6.
20
2

75

© 2006 Walch Publishing

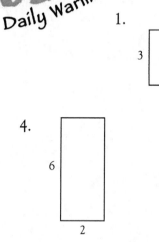

Angles

An **angle** is a way of measuring how two lines connected at a common end point separate from each other. The measure between the lines is made in degrees, and the degree symbol is °. The four basic kinds of angles are listed below.

Acute angle: any angle less than 90°

Obtuse angle: any angle between 90° and 180°

Right angle: any angle of exactly 90°

Straight angle: any angle of exactly 180°

Identify each angle below as one of the four kinds of angles listed above.

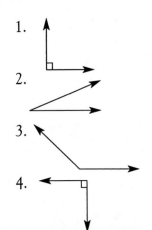

1.

2.

3.

4.

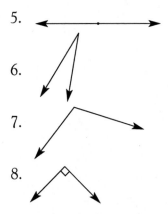

5.

6.

7.

8.

© 2006 Walch Publishing

Pairs of Angles

There are three pairs of angles that are found in geometry. Knowing what each pair looks like can help you solve problems that could be hard to do. The three pairs are listed below.

Complementary angles: The sum of the measures of two complementary angles is 90°.

Supplementary angles: The sum of the measures of two supplementary angles is 180°.

Adjacent angles: two angles that share a common side

Identify each angle below as one of the three kinds of angles listed above.

1.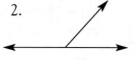

2.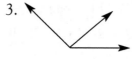

3.

4.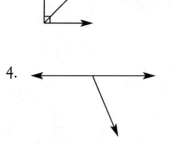

5.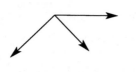

6.

77

© 2006 Walch Publishing

Vertical Angles

Vertical angles are found where any two lines intersect each other at an angle other than 180°, as in the image below.

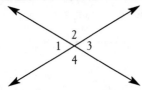

The opposite angles are vertical angles and are congruent, which means they have the same angle measure. The symbol for congruent is ≅. The symbol for angle is ∠. This means that ∠1 ≅ ∠3 and ∠2 ≅ ∠4. Any two angles that are side by side have a sum of 180°. From the picture, you can see that ∠1 + ∠2 = 180°, ∠2 + ∠3 = 180°, ∠3 + ∠4 = 180°, and ∠4 + ∠1 = 180°. This means if you know the measure of any one angle from the set of vertical angles, you can determine the measures of the other three.

Determine the missing angles in the problems below.

1.

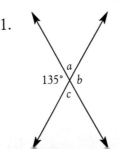

2.

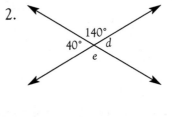

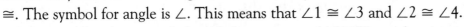

78

© 2006 Walch Publishing

Drawing Angles with Protractors

One tool that makes drawing angles much easier is a protractor. A protractor is marked with lines indicating different angles. By drawing a horizontal line and then a line that goes from the horizontal to a certain measurement on the curved part, you can draw almost any angle.

Use a protractor to draw the following angles.

1. 10°

2. 25°

3. 45°

4. 90°

5. 115°

6. 160°

79

© 2006 Walch Publishing

Parallel Lines and Angles

One unique situation in geometry is when a line crosses two parallel lines as shown below.

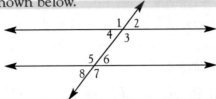

There is a lot of information in this image. The opposite vertical angles are congruent, which means they have the same angle measure. The symbol for congruent is ≅. The symbol for angle is ∠. This means that ∠1 ≅ ∠3, ∠2 ≅ ∠4, ∠5 ≅ ∠7, and ∠6 ≅ ∠8.

Alternate interior angles are congruent, so ∠3 ≅ ∠5. Alternate exterior angles are congruent, so ∠1 ≅ ∠7. Corresponding angles are congruent, so ∠1 ≅ ∠5.

Find the value of all the missing angles.

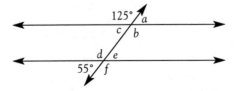

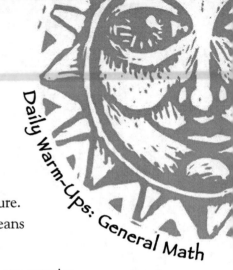

© 2006 Walch Publishing

Basic Polygons

There are a number of shapes that you will need to be able to draw to do basic geometry. Some of the basic polygons are shown below. The word *polygon* means "many sides." You can see that the number of sides is what determines the name of a polygon. You should also note that the sides of a polygon do not all have to be the same length.

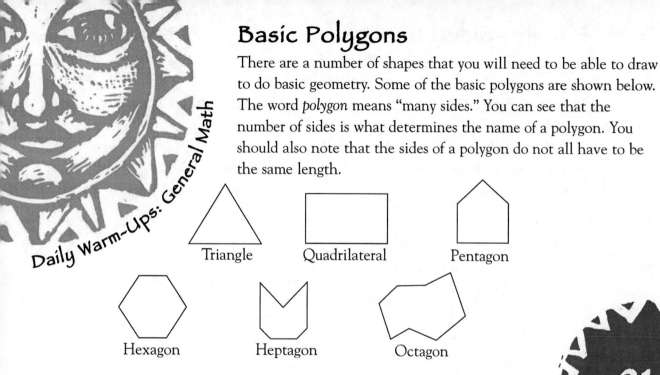

Triangle Quadrilateral Pentagon

Hexagon Heptagon Octagon

Draw one of each of the polygons shown above, but make it different from the example given.

81

© 2006 Walch Publishing

Basic Parallel-Sided Shapes

There are four basic shapes that have opposite sides that are parallel. The first two are the square and rectangle. Both have opposite sides that are parallel and congruent, and four angles that are right angles. Remember, all four sides of a square are congruent. For the parallelogram and the rhombus, both have opposite sides that are parallel and congruent, and angles at opposite corners are congruent as well. Remember, all four sides of a rhombus are congruent.

Identify the missing measures in the pictures below.

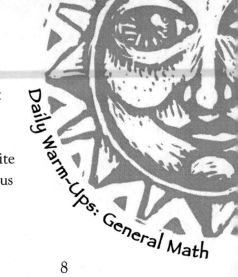

1.

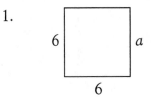

6   a

6

2.

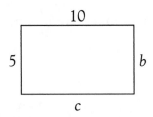

10

5   b

c

3.

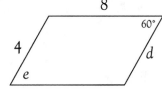

8

4   60°

d

e

4.

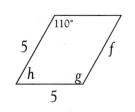

110°

5   f

h   g

5

82

Symmetry

A **line of symmetry** is a line that can be drawn through a figure or shape so that it cuts the figure into two pieces that are mirror images of each other. **Rotational symmetry** is when an object can be turned through less than 360° and come to a position such that it looks exactly as it did before it was rotated. The amount that the object is turned until it reaches such a position is called the **angle of rotation.**

Draw a line of symmetry through each object below that is symmetrical. Some objects may have more than one line.

1.

2.

3.

4.

5.

6.

7.

8. **A**

© 2006 Walch Publishing

Similar Figures

Similar figures are figures that are the same shape, but are not the same size. For example, look at the two squares below. They are only similar, and not identical.

The two shapes below are neither the same size nor the same shape.

They are not similar figures.

Identify the similar figures below.

1. 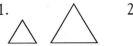   2.   3.   4.

5.   6.   7.

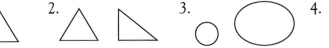

84

© 2006 Walch Publishing

Indirect Measurement

You can use similar figures to find the length, width, or height of an object that you cannot measure directly. When two figures are similar, the ratios of their corresponding sides are equal. You can write and solve a proportion to find the missing measurement. This is called **indirect measurement.**

Find the missing sides below by indirect measurement.

1.
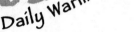

8

2

6

x

2.
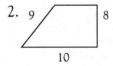

9   8

10

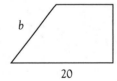

b   *a*

20

3.
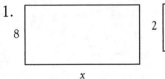

3   3

2

21   *x*

y

© 2006 Walch Publishing

Congruent Figures

Congruent figures are those figures or shapes that are the same size and the same shape. If the figures are the same shape but different sizes, then they are only similar. If the figures are of completely different sizes and shapes, then they are neither congruent nor similar.

Identify each shape below as similar, congruent, or neither.

1.

2.

3.

4.

5.

6.

© 2006 Walch Publishing

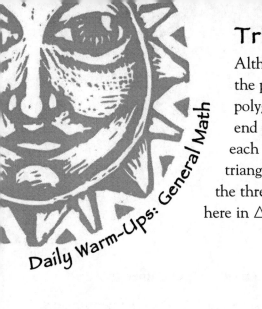

Triangles

Although you have seen triangles before, you probably don't know the precise definition from geometry. A **triangle** is a three-sided polygon formed when three line segments intersect only at their end points. A triangle has three interior angle measures, one at each vertex. The sum of the angles at the three vertices is 180°. A triangle is often described in text with the symbol △ followed by the three letters used to label the vertices of the triangle, as shown here in △ ABC. This is read as triangle A-B-C.

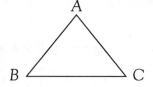

Daily Warm-Ups: General Math

Draw a triangle for each of the following descriptions.

1. △ XYZ, in which the distance from X to Y is the same as the distance from Y to Z.

2. △ QRS, in which the distance is the same between all three vertices.

3. △ JKL, in which the distance between all three vertices is different.

4. △ EFG, in which the distance from E to F is twice the distance from F to G.

87

© 2006 Walch Publishing

Triangles by Sides

There are three basic kinds of triangles as defined by the lengths of their sides. They are listed below.

Equilateral triangle: All three sides are the same length.

Isosceles triangle: At least two of the sides are the same length.

Scalene triangle: No two sides are the same length.

Use a ruler to complete the following problems.

1. Draw an equilateral triangle in which each side is 8 centimeters.

2. Draw an isosceles triangle in which the sides are 10, 10, and 5 centimeters.

3. Draw a scalene triangle in which the sides are 5, 10, and 12 centimeters.

4. Draw an equilateral triangle in which each side is 5.8 centimeters.

5. Draw an isosceles triangle in which the sides are 14, 10, and 14 centimeters.

6. Draw a scalene triangle in which the sides are 10, 14, and 12 centimeters.

© 2006 Walch Publishing

Triangles by Angles

There are three basic kinds of triangles as defined by the internal angle measures at their vertices. They are listed below.

Right triangle: a triangle that has one right angle

Acute triangle: a triangle in which all three angles are acute (less than 90°)

Obtuse triangle: a triangle in which one angle is obtuse (greater than 90°)

Use a protractor to complete the following problems.

1. Draw a right triangle that has one angle of 30°.

2. Draw an acute triangle that has one angle of 30°.

3. Draw an obtuse triangle that has one angle of 130°.

4. Draw a right triangle that has one angle of 40°.

5. Draw an acute triangle that has one angle of 40°.

6. Draw an obtuse triangle that has one angle of 150°.

89

Right Triangles

A right triangle always has one angle that has a measure of 90°. The side of the triangle that is opposite the 90° angle is called the **hypotenuse.** The other two sides of the triangle are called the **legs.** All the angles in any triangle have to add up to 180°. This means that in a right triangle, the two angles other than the right angle have to add up to 90°.

Find the missing angles in the right triangles below.

1.

2.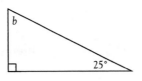

3.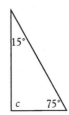

4.

5.

6.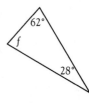

90

© 2006 Walch Publishing

Pythagorean Theorem

The **Pythagorean theorem** is a way of describing the relationship between the lengths of the three sides of a right triangle. The Greek philosopher Pythagoras discovered that for any right triangle, the sum of the squares of the lengths of the legs equals the square of the length of the hypotenuse. So for $\triangle ABC$, the sides are a, b, and c, as seen here.

For the triangle above, the Pythagorean theorem is $a^2 + b^2 = c^2$.

Find the length of the missing side for each right triangle below.

1.

2.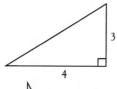

3.

4.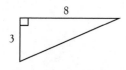

5.

6.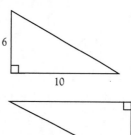

91

© 2006 Walch Publishing

Unique Right Triangles

There are two right triangles that have unique features that make them easy to use to solve problems. The first is the 45°/45°/90° triangle. The two legs a and b are always congruent, which means they have the same length. If you use the Pythagorean theorem, you can see that if $a = b$ then $a^2 + b^2 = c^2$ would become $a^2 + a^2 = c^2$, which is $2a^2 = c^2$. You only need to know the length of one of the legs to find the length of the hypotenuse.

The second triangle, the 30°/60°/90° triangle, has sides such that the hypotenuse is always twice the length of the shorter leg. In both triangles, these ratios allow you to find two missing sides if you know the length of any one side.

Find the lengths of the missing sides below.

92

1.

2.

3.

4.

© 2006 Walch Publishing

Daily Warm-Ups: General Math

Area of Parallelograms

The area of a rectangle can be found by multiplying the length times the width. In much the same way, we see the area of a parallelogram can be found by multiplying the length of the base of the parallelogram by the height. The **base** is any one of its sides, and the **height** is a line that runs perpendicularly from the base to the opposite side of the parallelogram as shown in the image below.

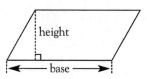

height

base

Find the areas of the following parallelograms.

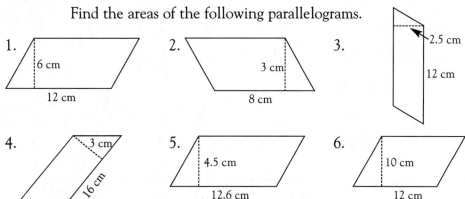

1.

6 cm

12 cm

2.

3 cm

8 cm

3.

2.5 cm

12 cm

4.

3 cm

16 cm

5.

4.5 cm

12.6 cm

6.

10 cm

12 cm

© 2006 Walch Publishing

93

Area of Triangles

A parallelogram is made of two identically sized triangles. If you run a diagonal between opposite corners of a parallelogram, you can see the two triangles as below.

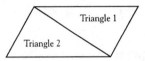

Triangle 1

Triangle 2

The area of a parallelogram can be found by multiplying the base times the height. A triangle can be thought of as half a parallelogram, so to find the area of a triangle, you can use the formula $A = \frac{1}{2}bh$, or area = one half base times height.

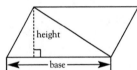

height

base

Find the area of each triangle below.

1.

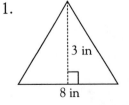

3 in

8 in

2.

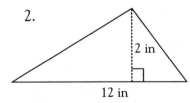

2 in

12 in

3.

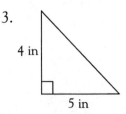

4 in

5 in

94

© 2006 Walch Publishing

Area of Trapezoids

A trapezoid is a special kind of quadrilateral in which only two of the sides are parallel. Here are some examples.

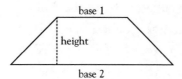

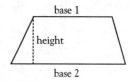

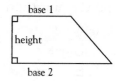

To find the area of a trapezoid, you can use a formula that is a combination of the formulas for area of a triangle and area of a parallelogram. The formula is area equals one half the height times the sum of the two bases, or $A = \frac{1}{2}h(b_1 + b_2)$.

Find the area of each trapezoid below.

1.

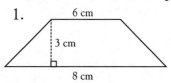

2.

3.

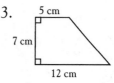

4.

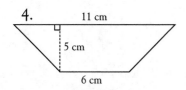

5.

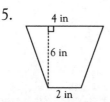

6.

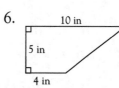

© 2006 Walch Publishing

95

Circumference of Circles

A **circle** is a set of all the points in a plane that are an equal distance from a point, called the **center.** The distance from the center point to the outside of the circle is called the **radius.** The distance across the circle that passes through the center is called the **diameter.** The distance around the perimeter of the circle is called the **circumference.** The ratio of the circumference to the diameter of a circle is called **pi** and is represented by the Greek letter π. The value of π is 3.14159265358979 . . . it just keeps going forever. We usually abbreviate pi and use 3.14 as the value in circle problems. The circumference of a circle is pi times the diameter or pi times twice the radius. The formulas are C = π*d* and C = 2π*r*.

Find the circumference of each circle below.

1.
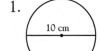
10 cm

2.

4 cm

3.

3 in

4.
12 in

5.

6.6 cm

6.

5.3 in

96

© 2006 Walch Publishing

Area of Circles

To find the area of a circle, you need to know the diameter or the radius. The formula for calculating the area of a circle is $A = \pi r^2$. The A is for area, the r is for radius, and π is equal to 3.14. There are many reasons why you would want to calculate the area of a circle. You might wonder whether or not it's a good deal to order two medium pizzas instead of one large pizza. For example, when you order a 16-inch pizza, you get a pizza that has a diameter of 16 inches. That means the radius is only 8 inches. So the area of the pizza is $A = \pi(8)2 = 201$ square inches.

Daily Warm-Ups: General Math

Which combination of pizzas gives you the most pizza for your money? Circle the letter of your answer.

a. 2 pizzas that cost $12 each and have a diameter of 20 inches

b. 3 pizzas that cost $8 each and have a diameter of 16 inches

c. 6 pizzas that cost $4 each and have a diameter of 12 inches

97

© 2006 Walch Publishing

Areas of Compound Figures

A **compound figure,** sometimes called a complex figure, is made up of other more common shapes. To find the area of a compound figure, you can break the figure into common shapes and calculate the areas of those shapes. When you have the individual areas, you can add them together to find the total area of the compound figure.

Find the total area for each compound figure below.

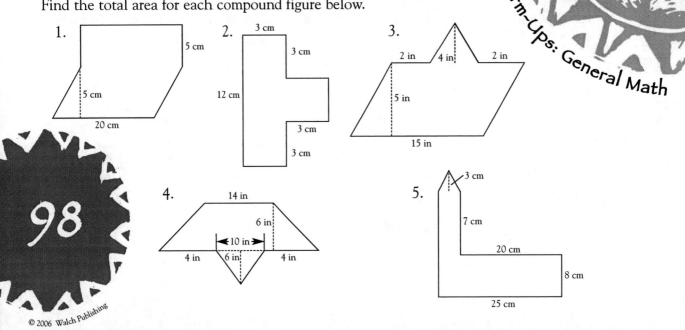

© 2006 Walch Publishing

Prisms and Pyramids

A **prism** is a three-dimensional shape that has two parallel congruent faces. A face is one of the flat surfaces of the shape. The edges are places where the faces meet and form lines. The vertices are the places where the lines meet to form corners. A **pyramid** is another kind of three-dimensional shape that has at least three faces that are triangles and has only one base. Both objects are three-dimensional and have faces, edges, and vertices.

Identify the number of faces, edges, and vertices for each of the following prisms and pyramids.

1.

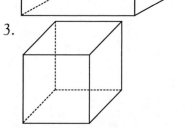

2.

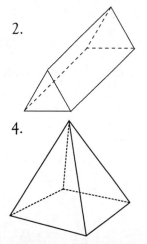

3.

4.

© 2006 Walch Publishing

99

Cylinders, Cones, and Spheres

A **cylinder** is an object that has three sides. Two of the sides are parallel and congruent surfaces that are both circles. A soda can would be a cylinder if the top and bottom were completely flat. A **cone** is an object that has two sides. One side of the cone is the base, and that base is a circle. An ice-cream cone and a traffic cone are two examples. A **sphere** is an object that only has one side and all the points on that side are of equal distance from the center. A ball or a marble is a good example of a sphere.

Answer the questions below.

1. Name three common items that are shaped like a cylinder.

2. Name three common items that are shaped like a cone.

3. Name three objects that are spheres.

100

© 2006 Walch Publishing

Drawing 3-D Shapes

Three-dimensional shapes have length, width, and height. When you look at an object such as a tissue box, you can see a large top, a long side, a short side, and you can even look at a corner or an edge. When you draw an object, you need to remember that there are parts you can't see. By turning an object, you can see it from more than one angle, and that can help you to draw it.

Draw the following objects from three angles other than the one shown.

1.
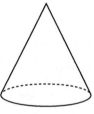

2.
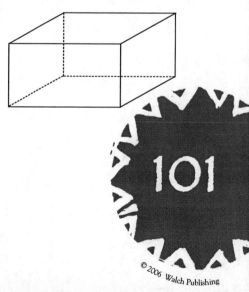

3.
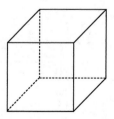

4.
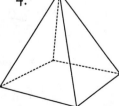

101

© 2006 Walch Publishing

Volume of a Rectangular Prism

The **volume** of an object is that amount of space it occupies. We take three-dimensional objects and determine how much space they occupy in a number of ways. One of those ways is to measure the object carefully and then to use formulas that mathematicians have determined for various shapes. Rectangular prisms have a length, a width, and a height. By multiplying these three measures together, you can find the volume. The formula is $V = l \cdot w \cdot h$. The units for volume are cubes, as in cubic centimeters or cubic inches.

Find the volume of each rectangular prism below.

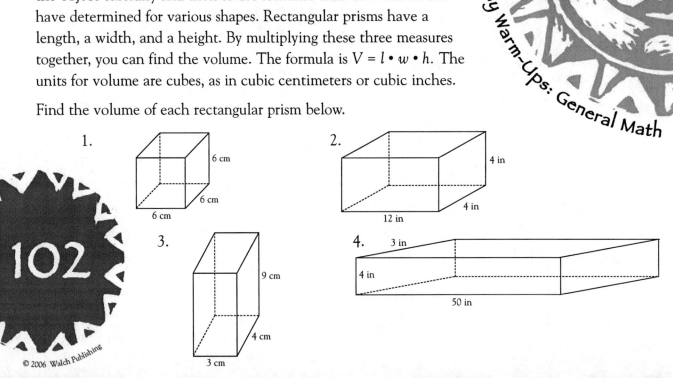

1. 6 cm, 6 cm, 6 cm

2. 4 in, 12 in, 4 in

3. 9 cm, 4 cm, 3 cm

4. 3 in, 4 in, 50 in

102

© 2006 Walch Publishing

Volume of a Triangular Prism

A **triangular prism** is a shape that has two congruent parallel triangles on the ends connected by three rectangles as shown below.

To find the volume of the triangular prism, you need to multiply the area of either triangle times the length of one of the rectangular faces. The formula is $V = \frac{1}{2} bhl$, where b is the base of the triangle, h is the height of the triangle, and l is the length of the rectangular face that isn't a side of the triangle.

Find the volume of each triangular prism below.

1.
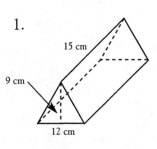
15 cm
9 cm
12 cm

2.
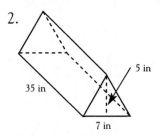
5 in
35 in
7 in

3.
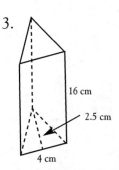
16 cm
2.5 cm
4 cm

103

© 2006 Walch Publishing

Volume of a Cylinder

The volume of a cylinder can actually be found by multiplying the area of the circle on the top or bottom of the cylinder by the height of the cylinder. This means that you need to know the radius of the circle on top or bottom of the cylinder and how tall or long the cylinder is. The formula for the volume of a cylinder is $V = \pi r^2 h$.

Find the volume of each cylinder below.

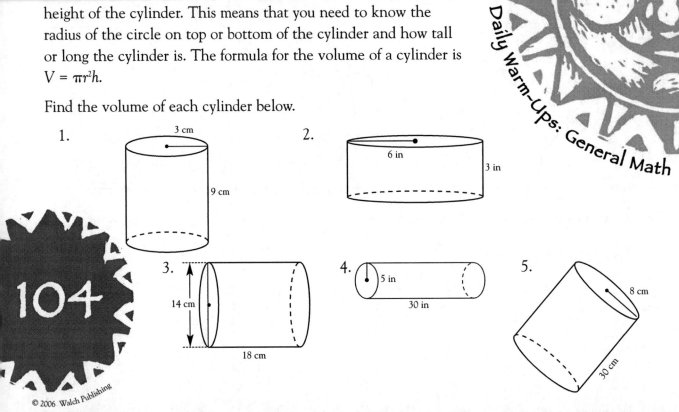

1. 3 cm 9 cm

2. 6 in 3 in

3. 14 cm 18 cm

4. 5 in 30 in

5. 8 cm 30 cm

104

© 2006 Walch Publishing

Volume of a Pyramid

A pyramid is another kind of three-dimensional shape that has at least three faces that are triangles and only one base. The base can be a triangle, a square, a pentagon, or any other polygon. To calculate the volume of a pyramid, you have to be able to calculate the volume of the base. Any shape that you can calculate the area of can be used as the base. The height of the pyramid is a line that runs from the peak of the pyramid down to the base so that it forms a right angle with the base. The formula for the volume of a pyramid is $V = \frac{1}{3}Bh$. The B in the formula stands for area of the base.

Calculate the volume of each pyramid below.

1.

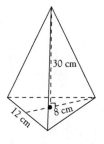

30 cm

12 cm   8 cm

2.

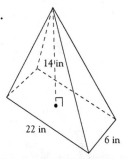

14 in

22 in   6 in

3.

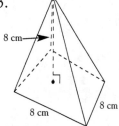

8 cm →

8 cm   8 cm

105

© 2006 Walch Publishing

Volume of a Cone

A cone is an object that has two sides. One side of the cone is the base, and that base is a circle. The cone is a little unusual because it does not have any edges or a vertex. The basic formula for the volume of a cone is the same as the formula for the volume of a pyramid, $V = \frac{1}{3}Bh$. However, in this case, the area represented by the letter B is the area of a circle, or $A = \pi r^2$. If we use πr^2 for the letter B, the formula is $V = \frac{1}{3}\pi r^2 h$.

Find the volume of each cone below.

106

1.
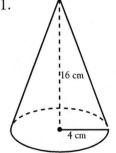
16 cm
4 cm

2.
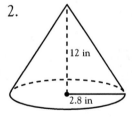
12 in
2.8 in

3.
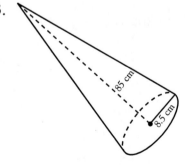
85 cm
8.5 cm

© 2006 Walch Publishing

Volume of a Sphere

A sphere is an object that only has one side. All the points on that side are of equal distance from the center of the sphere. There are spheres all around us. Baseballs, softballs, tennis balls, marbles, ball bearings, and many other objects are spheres. To calculate the volume of a sphere, you need to know the radius of the sphere and use the formula $V = \frac{4}{3} \pi r^3$. In the formula, r is the radius. Be sure to remember that the radius isn't squared, it's cubed.

Find the volume of each sphere below.

1.

3 in

2.

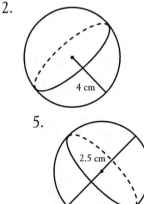

4 cm

3.

12 cm

4.

16 in

5.

2.5 cm

107

© 2006 Walch Publishing

Surface Area of a Rectangular Prism

The surface area of a rectangular prism can be found by adding the surface areas of all the individual faces of the prism. For example, if you take a rectangular prism and unfold it so that all the surfaces can be seen, you see that there are actually six rectangles. Opposite rectangles are equal in size.

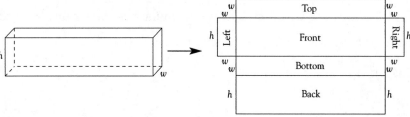

You can see that $l \cdot w$ occurs twice, $h \cdot w$ occurs twice, and $l \cdot h$ occurs twice. This means the formula for surface area of a rectangular prism is $S = 2(lw + hw + lh)$.

Find the surface area of each rectangular prism below.

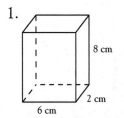

1.
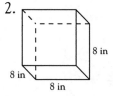
8 cm
2 cm
6 cm

2.
8 in
8 in
8 in

3.
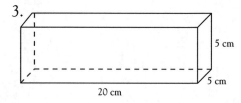
5 cm
5 cm
20 cm

© 2006 Walch Publishing

Surface Area of an Equilateral Triangular Prism

The surface area of a triangular prism can be found by adding the surface areas of all the individual faces of the prism. For example, if you take a triangular prism and unfold it so that all the surfaces can be seen, you see there are two triangles and three rectangles.

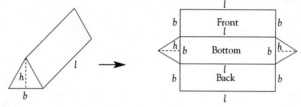

To find the surface area, add the areas of the two triangles to the areas of the three rectangles. The formula for the surface area of an equilateral triangular prism is $S = bh + 3lb$.

Find the surface area of each equilateral triangular prism below.

1.

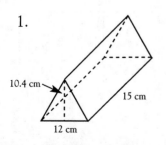

10.4 cm

15 cm

12 cm

2.

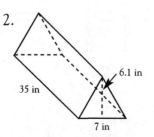

35 in

6.1 in

7 in

3.

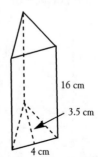

16 cm

3.5 cm

4 cm

109

© 2006 Walch Publishing

Surface Area of a Cylinder

A cylinder is an object with two sides that are parallel and congruent surfaces that are both circles. The diagram shows how two circles and a rectangle form a cylinder.

 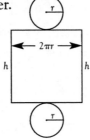

To find the surface area of a cylinder, you need to add the areas of the two circles to the area of the rectangle. The formula for the surface area of a cylinder is $S = 2\pi r^2 + 2\pi rh$, where h is the height of the cylinder.

Find the surface area of each cylinder below.

110

1.
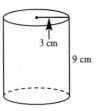
3 cm
9 cm

2.
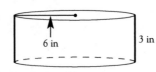
6 in
3 in

3.
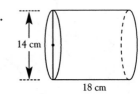
14 cm
18 cm

© 2006 Walch Publishing

Surface Area of a Pyramid

A pyramid is another kind of three-dimensional shape that has only one base and at least three faces that are triangles. The base can be a triangle, a square, a pentagon, or any other polygon. There is no simple formula for the surface area of a pyramid because the shape of the base could be any one of a number of different shapes. The total surface area is found by carefully counting all the faces and then calculating the surface area for each one. When this is done, you must add all the individual surface areas to find the total surface area for the whole pyramid.

Find the surface area of each pyramid below.

1.

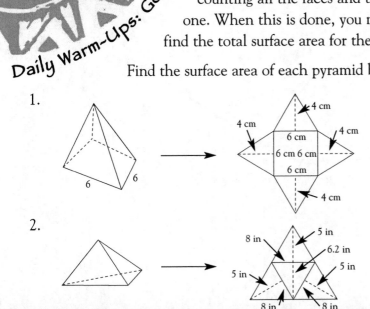

4 cm
4 cm
4 cm
6 cm
6 cm 6 cm
6 cm
4 cm

2.

8 in   5 in
6.2 in
5 in   5 in
8 in   8 in

© 2006 Walch Publishing

Surface Area of a Cone

A cone is an object that has two sides. One side of the cone is the base, and that base is a circle. The cone is a little unusual because it doesn't have any edges or a vertex. To find the surface area of a cone, you need to find the surface area of the circle that makes up the base (πr^2) and add it to the surface area of the slanted portion (πrl). The formula is $S = \pi r^2 + \pi rl$. The l is the distance along the slant of the cone.

Find the surface area of each cone below.

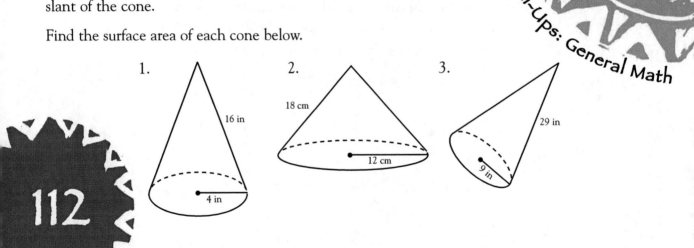

1. 16 in   4 in

2. 18 cm   12 cm

3. 29 in   9 in

112

© 2006 Walch Publishing

Surface Area of a Sphere

A sphere is an object that only has one side. All the points on that side are of equal distance from the center of the sphere. There are spheres all around us. Both bowling balls and racquetballs are kinds of spheres. To find the surface area of a sphere, you can use the formula $S = 4\pi r^2$. In the formula, the r stands for the radius of a sphere, which is the distance from the center of the sphere out to a point on the surface.

Find the surface area of each sphere below.

1.
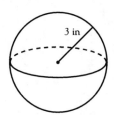
3 in

2.
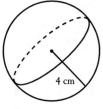
4 cm

3.
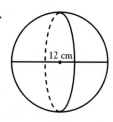
12 cm

4.
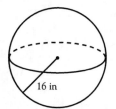
16 in

5.
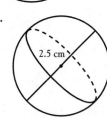
2.5 cm

113

© 2006 Walch Publishing

Tessellations

A **tessellation** is a repeating pattern made of polygons that fit together in such a way that they do not overlap and do not have any gaps between them. A common tessellation that most people have seen is a checkerboard. Below is a tessellation made from hexagons.

Note: At any point where the vertices meet, the sum of the angles always has to be 360°.

Follow the directions below.

1. Draw a tessellation that uses rectangles.

2. Draw a tessellation that uses triangles.

3. Draw a tessellation that uses parallelograms.

4. Draw a tessellation that does not use regular polygons.

114

© 2006 Walch Publishing

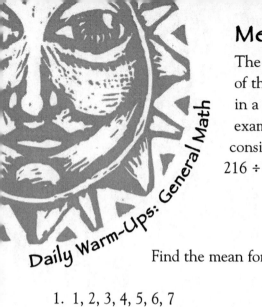

Mean

The mean of a group of numbers is often referred to as the average of the group of numbers. The **mean** is the sum of all the numbers in a group divided by the number of numbers in the group. For example, the group of numbers 12, 13, 69, 34, 9, 2, 50, and 27 consists of eight numbers. The sum of those numbers is 216. $216 \div 8 = 27$. The mean of the group of numbers is 27.

Find the mean for each group of numbers below.

1. 1, 2, 3, 4, 5, 6, 7

2. 5, 10, 15, 20

3. 100, 200, 150, 1000, 44

4. 3, 6, 9, 12, 15, 18, 21

5. 6, 8

6. 55, 77, 99

7. 300, 1, 66.5, 12.8

8. 9, 9, 9, 9, 9

9. 21, 34, 54, 56, 78

10. 79, 10, 20, 45, 606

115

© 2006 Walch Publishing

Median

The **median** of a group of numbers can be found by ordering the numbers, usually from smallest to largest, and then finding the middle number. For example, if you put the numbers 1, 12, 3, 10, 8, 9, 4, 6, and 7 in order, then you would have 1, 3, 4, 6, 7, 8, 9, 10, 12. The central number of the ordered list is 7, so 7 is the median of the list. If the list doesn't have an exact middle, you can find the mean of the two central numbers. For example, if you put the numbers 1, 5, 4, 7, 9, and 8 in order, then you would have 1, 4, 5, 7, 8, 9. The central numbers are 5 and 7, so you would find their mean, which is 6. This means that 6 is the median of the list.

116

Find the median of each list of numbers below.

1. 2, 3, 9, 7, 5, 6, 4

2. 1, 1, 25, 17, 16, 9, 3

3. 150, 562, 341, 999, 168, 227

4. 21, 22, 29, 28, 23, 24, 27, 26

5. 1, 99, 2, 98, 3, 97, 4, 96, 5, 95

6. 11, 80, 14, 17, 77, 74, 20

7. 121, 99, 149, 71, 66, 80, 134, 11

8. 2, 2, 3, 3, 7, 8, 9, 5, 4, 3, 3, 6, 7

© 2006 Walch Publishing

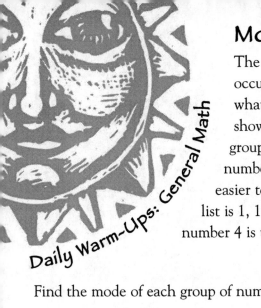

Mode

The **mode** of a group of numbers is the number or numbers that occur most often in the group. For example, if you wanted to know what part of the population watches a certain kind of television show, such as cartoons, you would collect all the ages and then group them. For example, say you gather the following group of numbers: 1, 1, 5, 4, 7, 9, 8, 4, 3, 2, 2, 2, 4, 5, 6, 4, 7. It is probably easier to find the mode if you put the numbers in order. The ordered list is 1, 1, 2, 2, 2, 3, 4, 4, 4, 4, 5, 5, 6, 7, 7, 8, 9. It is now clear that the number 4 is the mode of the group.

Find the mode of each group of numbers below.

1. 1, 1, 2, 2, 3, 3, 7, 8, 1, 3, 2, 5, 4, 6, 6, 1, 2, 1

2. 5, 10, 15, 10, 15, 10, 5, 5, 5, 20, 15, 5

3. 123, 121, 122, 125, 124, 123, 121, 125, 123, 121, 125, 123

4. 8, 9, 7, 5, 8, 6, 7, 5, 8, 7, 7, 9, 8, 9, 9, 6, 5, 6, 8

5. 17, 18, 21, 20, 17, 17, 19, 19, 18, 20, 21, 21, 17, 18, 19, 17

117

© 2006 Walch Publishing

Stem-and-Leaf Plots

A **stem-and-leaf plot** is another way of organizing a large collection of data. Once the data is collected, it becomes easier to see things such as which values are largest or smallest, or which values occur most often. For example, if you went to the movies and collected all the ages of the people attending one movie, the data could be organized as below.

5 | 1 6 6 can be read as 51, 56, 56 and indicates that one person who was 51 and two people who were 56 attended the movie.

Complete the problem below.

Stem	Leaf
0	899
1	01234555567777889
2	01357799
3	33469
4	245
5	166

118

A car company attempting to appeal to young consumers (ages 18–30) wants to know how successful they have been. They decide to keep track of the ages of all the people who came to test-drive their new car. Below are the numbers they collected in a week. Organize the numbers into a stem-and-leaf plot, and decide whether the company was successful or not.

21, 23, 70, 45, 34, 17, 44, 21, 67, 51, 20, 20, 31, 22, 18, 19, 35, 39, 21, 21, 34, 45, 73, 40, 21, 17, 17, 21, 54, 18, 18, 19, 34, 33

© 2006 Walch Publishing

Variation

Variation is a way of describing how data is distributed. One of the ways to look at the range of a set of numbers is to split the data into quartiles. To determine the quartiles for a group of numbers, you must first order the numbers from largest to smallest and then split them into four parts, with each part containing an equal number of numbers. This is done by determining the median of the whole group, which splits the group into two halves. The median is found for the lower half of the data and is called the **lower quartile.** The median is also found for the upper half of the data and is called the **upper quartile.**

Order each group of numbers below, and identify the median, the upper quartile, and lower quartile.

1. 1, 8, 7, 3, 6, 5, 4, 4, 4, 5, 3, 7, 8, 4, 3

2. 23, 24, 30, 23, 33, 23, 22, 40, 35, 35, 24, 50, 41, 44, 45

3. 175, 155, 172, 164, 169, 171, 158, 155, 174, 180, 177, 152, 156, 166, 178

4. 1.8, 2.1, 2.0, 1.7, 1.7, 1.9, 1.9, 1.8, 2.0, 2.1, 2.1, 1.7, 1.8, 1.9, 1.7

119

© 2006 Walch Publishing

Interquartile Range

Range is one way to measure variation, and it tells us how the numbers from some data are spread out. **Range** is the difference between the largest number in a group and the smallest number in a group. **Interquartile range** is the difference between the value of the upper quartile and lower quartile for a set of numbers. For example, the set of numbers 9, 1, 2, 3, 7, 8, 1, 3, 2, 5, 4, 6, 6, 1, 2, 1 is ordered to read 1, 1, 1, 1, 2, 2, 2, 3, 3, 4, 5, 6, 6, 7, 8, 9. The median is 3, and the group is split between the two 3s. The value of the lower quartile is 1.5, and the value of the upper quartile is 6. The interquartile range is 6 − 1.5 = 4.5.

Find the interquartile range for each set of numbers below.

120

1. 4, 3, 6, 2, 7, 8, 7, 7, 7, 6, 5, 3, 2, 5, 4

2. 66, 61, 69, 68, 65, 64, 67, 63, 64, 68, 66, 65, 68, 69, 71

3. 210, 200, 211, 193, 194, 209, 202, 191, 203, 190, 188, 216, 179, 185, 195

4. 1.5, 8.5, 7.5, 3.5, 6.5, 5.5, 4.5, 4.5, 4.5, 5.5, 3.5, 7.5, 8.5, 4.5, 3.5

5. 1.50, 1.55, 1.72, 1.64, 1.69, 1.71, 1.58, 1.55, 1.74, 1.80, 1.77, 1.52, 1.56, 1.66, 1.78

© 2006 Walch Publishing

Outliers

The numbers in a group may all be very close in value, but sometimes numbers seem far outside the range of the numbers you are working with. These numbers are called **outliers.** To find an outlier, you need to first find the lower quartile and upper quartile for a set of numbers. If the lower quartile is 21 and the upper quartile is 30, you can find the outliers by following these steps.

1. Find the difference between the upper and lower quartile: 30 – 21 = 9.

2. Find 1.5 times the difference: 1.5 × 9 = 13.5.

3. To find outliers below your range, take the lower quartile and subtract 13.5: 21 – 13.5 = 7.5.

4. To find outliers above your range, take the upper quartile and add 13.5: 30 + 13.5 = 43.5.

Any numbers below 7.5 or above 43.5 are outliers.

Find the values of outliers for each set of numbers below.

1. 4, 3, 6, 2, 7, 8, 7, 7, 7, 6, 5, 3, 2, 5, 4

2. 150, 155, 172, 164, 169, 171, 158, 155, 174, 180, 177, 152, 156, 166, 178

3. 2.3, 2.4, 3.0, 2.3, 3.3, 2.3, 2.2, 4.0, 3.5, 3.5, 2.4, 5.0, 4.1, 4.4, 4.5

121

© 2006 Walch Publishing

Box-and-Whisker Plots

A **box-and-whisker plot** is a way of graphing a set of numbers on a number line in a manner that includes the median, the upper and lower quartiles, and any points that may be outliers.

1. Draw a number line that has a range that will include the minimum and maximum values from your set of numbers.

2. Mark the median.

3. Mark the least value and next-to-least value that are not outliers.

4. Mark the greatest value and next-to-greatest value that are not outliers.

5. Draw a box around the lower quartile from the median to the next-to-least value.

6. Draw a line from the box down to the least value that is not an outlier.

7. Draw a box around the upper quartile from the median to the next-to-greatest value.

8. Draw a line from the box up to the greatest value that is not an outlier.

Follow the steps above. Draw a box-and-whisker plot for each set below.

1. 65, 66, 61, 69, 68, 50, 65, 64, 67, 63, 64, 68, 66, 85, 65, 68, 69, 71

2. 190, 210, 200, 211, 193, 194, 209, 202, 191, 203, 190, 188, 212, 179, 185, 195

122

© 2006 Walch Publishing

Arithmetic Sequences

An **arithmetic sequence** is one in which the difference between any two consecutive terms is the same. In this case, a term means a number. For example, the sequence 2, 4, 6, 8, 10, 12, . . . increases by 2 each time. The number 2 is called the common difference, and any number in the list is one of the terms of the sequence. In the sequence $\frac{1}{2}$, 1, $1\frac{1}{2}$, 2, $2\frac{1}{2}$, 3, $3\frac{1}{2}$, 4, $4\frac{1}{2}$, 5, $5\frac{1}{2}$, 6, $6\frac{1}{2}$, . . . the common difference is $\frac{1}{2}$. Not all arithmetic sequences increase in value. For example, the sequence 25, 20, 15, 10, 5, 0, –5, –10, –15, –20, –25, . . . decreases by 5 each time. Its common difference is –5.

Find the common difference of each arithmetic sequence.

1. 4, 8, 12, 16, 20, . . .

2. 5, 7.5, 10, 12.5, 15, 17.5, . . .

3. 23, 27, 31, 35, 39, . . .

4. 78, 74, 70, 66, 62, 58, . . .

Complete each arithmetic sequence.

5. 18, 24, 30, _____, _____, _____, . . .

6. 121, 110, 99, _____, _____, _____, . . .

7. 56, 98, 140, _____, _____, _____, . . .

8. 14, _____, 28, _____, 42, _____, . . .

© 2006 Walch Publishing

Geometric Sequences

A **geometric sequence** is one in which each term can be found by multiplying the preceding number by a constant. For example, in the sequence 2, 4, 8, 16, 32, 64, 128, . . . each term was multiplied by 2 to get the next term. In this case, the number 2 is called the common ratio. In the sequence 3, 9, 27, 81, 243, 729, 2187, . . . the common ratio is 3. You can see that geometric sequences grow very rapidly when the common ratio is not very large. A geometric sequence does not always grow larger, however. For example, in the sequence 1000, 500, 250, 125, 62.5, 31.25, 15.625, 7.1825, . . . the common ratio is 0.5, and the sequence grows smaller rapidly.

Find the common ratio of each geometric sequence.

1. 80, 40, 20, 10, 5, . . .

2. 7, –21, 63, –189, 567, –1701, 5103, . . .

124

Complete each geometric sequence.

3. 64, _____, 256, _____, 1024, _____, 4096, 8192, . . .

4. 66, 1452, 31944, _____, 15460896, . . .

© 2006 Walch Publishing

Coordinate Plane

The **coordinate plane** is a grid made up of two number lines that intersect at their zero points. These two number lines intersect at a 90° angle and divide the grid into four quadrants. The horizontal line is called the *x*-axis, and the vertical line is called the *y*-axis. The point of intersection is called the **origin.** The terms mentioned here are shown in the diagram of a coordinate plane on the right.

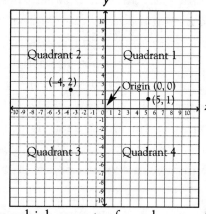

Points can be located through the use of ordered pairs, which are sets of numbers that give the location of a point with the *x*-axis coordinate first and the *y*-axis coordinate second. The origin is at (0, 0). Two sample points are on the coordinate plane above. The two points are (5, 1) and (– 4, 2).

Make a coordinate plane on a sheet of graph paper, and plot the points below.

1. (1, 6)   2. (3, –5)   3. (2, –2)   4. (–1, 4)   5. (–3, –8)

125

© 2006 Walch Publishing

Distance Between Points

If two points are on the same horizontal or vertical line on the coordinate plane, you can subtract to find the distance between them. For example, two points (10, 10) and (10, 4) are on the same vertical line. The distance between them is 10 − 4 = 6. If two points are not on the same vertical line, however, you can use the Pythagorean theorem to find the distance between them. First, plot the two points and make a right triangle that has the points on the two vertices that are not opposite the right angle. Then find the length of sides a and b and use the Pythagorean theorem to find the length of the hypotenuse. The process has been completed for the points (1, 5) and (4, 1) to the right.

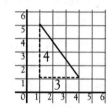

126

The triangle has a length of 4 for one leg and 3 for the other. The Pythagorean theorem gives us $4^2 + 3^2 = c^2$, so $c^2 = 25$ and $c = 5$.

Find the length between each set of points below.

1. (5, 5) (3, 3)
2. (25, 1) (25, 8)
3. (8, 9) (12, 20)
4. (−2, 6) (−10, 8)
5. (−10, 4) (−1, 6)
6. (2, 2) (−6, −6)

© 2006 Walch Publishing

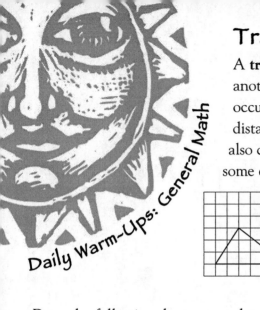

Translations

A **translation** is made when a figure is moved from one position to another without turning. On a coordinate plane, a translation occurs when all the points from a figure are moved the same distance and direction from the original figure. The new figure is also congruent to the first figure and is not turned at all. Below are some examples of translations.

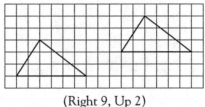

(Right 9, Up 2)

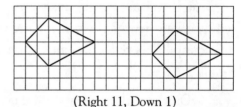

(Right 11, Down 1)

Draw the following shapes on a sheet of graph paper, and perform the translations as described.

1. 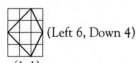 (Left 6, Down 4)

(1, 1)

2. 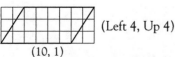 (Left 4, Up 4)

(10, 1)

3. 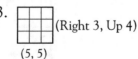 (Right 3, Up 4)

(5, 5)

4. 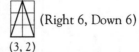 (Right 6, Down 6)

(3, 2)

127

© 2006 Walch Publishing

Reflections

A **reflection** is made when a mirror image is produced by flipping a figure over a line. Simple translations on the coordinate plane are usually made across the *x*-axis or across the *y*-axis. Every point on the reflection is the same distance from the line of reflection as the original point was from the line of reflection. The image produced will be congruent to the original image, but it will not be oriented the same. Here is an example of a reflection about the *y*-axis.

Draw a reflection of the image across the *x*-axis, then across the *y*-axis.

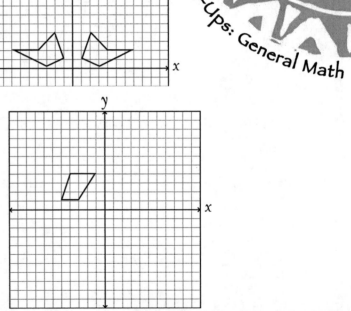

128

© 2006 Walch Publishing

Rotations

A **rotation** is formed when a figure is turned around a fixed point called the center of rotation. A point on the rotation and the corresponding point on the original figure are the same distance from the center of rotation. The images are congruent, and the orientations of the original and the rotation are the same. Here is an example of a rotation with the origin of the coordinate plane as the center of rotation. The figure on the right is the original.

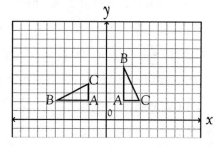

If you check ∠AOA', you will see that it is 90°, as are ∠BOB' and ∠COC'. The angle of rotation has to be the same for all points on the original and the rotation, but it does not have to be 90°. Other angles will also work.

Rotate the image to the right 90° three times so that it appears in all four quadrants. Use the origin as your center of rotation.

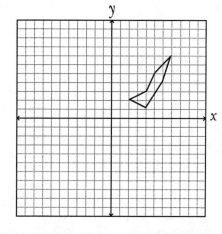

© 2006 Walch Publishing

Ratios

A **ratio** is a way of comparing two numbers by division. If you have a group of books and there are 10 hardcovers and 14 paperbacks, the ratio of hardcovers to paperbacks is written as 10:14. The ratio can also be written as $\frac{10}{14}$. Remember, a fraction such as $\frac{10}{14}$ can be reduced to $\frac{5}{7}$. The ratio of hardcover books out of the total would be 10:24, which is read as "ten to twenty-four" or "ten out of twenty-four."

Complete the following problems.

130

1. Write a ratio for the number of your fingers to the number of your hands.

2. Write a ratio for the number of cars your family owns to the number of people in your family.

3. Write the ratio $\frac{21}{28}$ in simplest form.

4. Write the ratio $\frac{100}{600}$ in simplest form.

5. Write a ratio for the number of centimeters in a meter.

6. Write a ratio for the number of inches in a yard.

© 2006 Walch Publishing

Rates

A **rate** is a kind of ratio that is used to compare two quantities with different units. The mileage of a car is often described as 30 miles to the gallon, a ratio of miles to gallons. As a fraction, this ratio would be $\frac{30 \text{ miles}}{\text{gallon}}$. Another ratio might be a measure of car speed, such as $\frac{80 \text{ kilometers}}{\text{hour}}$. Not all rates that are ratios need distance or time in them. At the supermarket, you may see the price of fruits or vegetables listed in dollars per pound or some similar measure. For example, grapes might be $\frac{\$2.50}{\text{pound}}$, or $\frac{\$3.00}{\text{bunch}}$.

Complete the following problems.

1. Think of a common rate you might see on a speed limit sign.

2. For each of the following, circle the rate that indicates the better buy.

   a. 4 batteries for $3.00 or 8 batteries for $6.25

   b. $3.50 for a dozen eggs or 30¢ per egg

   c. 75¢ for a can of soda, or $2.99 for a six-pack of soda

   d. a 32-ounce bag of chips for $2.89 or a 64-ounce bag of chips for $5.78

131

© 2006 Walch Publishing

Simple Probability

A **probability** is a kind of ratio that compares the number of favorable outcomes to the total number of possible outcomes. Probability is the chance that some particular event will occur. For example, if you flip a quarter to get heads or tails, there are two possible outcomes. If you are trying to get heads, then there is one favorable outcome out of two total outcomes, or $\frac{1}{2}$, which represents a one-in-two chance. This ratio also produces a percent. The chance of getting heads when flipping a quarter is 50%. If you had 4 blue socks and 6 red socks in a bag, then the chances of choosing a blue sock would be 4 out of the total of 10 socks, or $\frac{4}{10}$. This is the same as $\frac{2}{5}$, or 40%.

Complete the problems below.

1. What is the probability of rolling a 4 on a 6-sided die?

2. What is the probability of getting a blue marble from a bag that has 12 blue marbles and 23 green marbles?

3. What is the probability of getting a girl's name if you put the names of all the students in your class into a hat and chose one at random?

132

© 2006 Walch Publishing

Counting Outcomes

An **outcome** is one single event that can occur when using some item to determine probability. For example, when rolling numbers on a 6-sided die, the outcomes are 1, 2, 3, 4, 5, and 6. There are six possible outcomes. When looking at any event for probability, it is necessary to be able to determine the total number of outcomes. For example, if you were choosing names at random from a box for a raffle, the number of outcomes would be the same as the number of tickets that were put in the box. In a lottery, however, the probability of having a winning ticket depends not on the number sold, but on all the possible combinations of numbers than can be drawn.

Solve the following problems.

1. What would be the total number of outcomes if you chose a letter from the alphabet at random?

2. What would be the total number of outcomes if you rolled a 12-sided die?

3. What would be the total number of outcomes if you chose a card from a regular 52-card deck of playing cards?

4. What would be the total number of outcomes if you printed 500 tickets for a raffle, but only sold 475?

133

© 2006 Walch Publishing

Tree Diagrams

A **tree diagram** is a way of showing all the possible outcomes for some series of events, or for combinations of outcomes. For example, if you were flipping a coin to get heads or tails, a tree diagram of the possible outcomes would look like this:

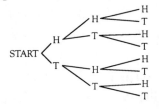

As you can see, the top branch indicates flipping heads 3 times in a row. The bottom branch shows flipping tails 3 times in a row. The center branches show all the possible combinations of heads and tails that could have been produced as well. At each branching point, all the possible outcomes split off to produce more branches and more outcomes.

Create a tree diagram for all the combinations you could get if you were buying a hat and the hat could

  be from one of two teams: bears or lions

  be in one of two colors: brown or gold

  have trim in one of two colors: white or black

© 2006 Walch Publishing

Probability Areas

Probability can be represented as a percent. These percents can be used to define an area that represents both the desired outcome for a probability event as well as the total number of outcomes. For example, on a 10 × 10 grid with 100 boxes, shading in 30 of the boxes would represent a 30% chance of a desired event occurring. The 70 boxes that were not shaded would represent the likelihood that the event would not take place.

A standard NFL football field is 300 feet long between goal lines and 160 feet wide. This means it is 3600 inches × 1920 inches and has an area of 6,912,000 square inches. If your odds of winning the lottery are 1 in 7 million, winning would be like flying over a football field in a helicopter and trying to hit an area a little smaller than a single square inch with a dart, with your eyes closed!

Complete the following problems.

1. Draw a probability area that shows a 50% chance.

2. Draw a probability area that shows a 25% chance.

3. If you have a dartboard that is 4 boxes by 4 boxes, what are the odds of the dart landing in any one square?

4. Draw a probability area that shows a 20% chance, but do not use squares.

135

© 2006 Walch Publishing

Fundamental Counting Principle

The **fundamental counting principle** is a way of determining all of the possible outcomes of an event by using multiplication instead of counting individual outcomes or using a tree diagram. The fundamental counting principle says that if an event A can happen *a* ways and if followed by an event B that can happen *b* ways, then event A followed by event B can happen *a* × *b* ways. For example, imagine you flip a coin four times and want to see the number of possible combinations that can happen in a certain order. There are 2 outcomes for each flip, so the total number of outcomes is 2 × 2 × 2 × 2 = 16 outcomes.

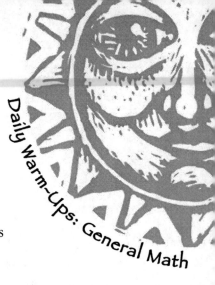

Complete the following problems.

136

1. How many outcomes are there if you roll a 6-sided die 3 times?

2. How many outcomes are there if you choose a card from a 52-card deck of playing cards, replace the card and reshuffle, and draw again?

3. If you need to come up with a password that has 3 letters and 2 numbers, how many combinations are there?

4. How many combinations are there for a password if you could use 4 letters and 4 numbers?

© 2006 Walch Publishing

Permutations

A **permutation** is a set of outcomes that appear as an arrangement or a list in which the order matters. For example, say you are deciding the order that three speakers, Rico, Gem, and Kimber, could walk out onstage. The combinations are as follows:

Rico, Gem, Kimber	Kimber, Gem, Rico
Rico, Kimber, Gem	Gem, Rico, Kimber
Kimber, Rico, Gem	Gem, Kimber, Rico

There were 3 choices of who could walk out first, but once the first person is out, there are only 2 people left to go second. After the second person goes out, there is only one person left. The permutations are 3 × 2 × 1 = 6.

Complete the problems below.

1. How many permutations are there for drawing 3 cards from a deck of 52 playing cards? (Remember, once you draw a card, there are only 51 left.)

2. How many permutations are there for the first, second, and third place finishers in a road race with 35 people in it?

3. How many permutations are there for you and 5 friends to line up single file?

137

© 2006 Walch Publishing

Combinations

A **combination** is a set of outcomes that appear as an arrangement or a list in which the order does not matter. For example, if you wanted to get pizzas that had all the combinations of pepperoni, ham, and sausage, taken 2 at a time, the combinations are

a. pepperoni, ham

d. ham, sausage

b. pepperoni, sausage

e. sausage, pepperoni

c. ham, pepperoni

f. sausage, ham

If you look, you can see that choices a and c are the same, choices b and e are the same, and choices d and f are the same. There are only three different combinations that you can make.

138

Complete the following problems.

1. How many combinations, taken 3 at a time, can you make for a cake topping from frosting, coconut, pecans, walnuts, and chocolate chips?

2. How many combinations, taken 2 at a time, can you make for a gift basket from little bags of chocolates, caramels, peppermints, butterscotch candies, and peanut clusters?

© 2006 Walch Publishing

Theoretical Probability

The **theoretical probability** of an event is determined based on a set of known facts or known properties of an event or an object. For example, each face on a 6-sided die has an equal chance of appearing when the die is rolled. The probability of any one face appearing is 1 in 6. Theoretical probabilities are seldom determined by experimentation. For example, if the odds of drawing the winning number in a lottery are 1 in 30 million, the odds were not determined by drawing millions of slips. By the same token, a quarter only has 2 faces, and as such, can only reveal one or the other. The odds of a coin showing heads are 1 in 2.

Complete the following problems.

1. What is the theoretical probability of rolling double 6s with a pair of dice?

2. What is the theoretical probability of drawing the king of hearts from a deck of 52 cards?

3. What is the theoretical probability of drawing the kings of hearts from a deck of 52 cards, and then drawing the queen of hearts?

4. If you were to flip a quarter 4 times, what is the chance that you would get tails all 4 times?

139

© 2006 Walch Publishing

Experimental Probability

Experimental probabilities are usually determined by collecting data. For example, if you and a friend were to flip a quarter 100 times and keep track of how many times the quarter landed showing heads and showing tails, you would probably expect it to be 50/50. However, you might find that you got tails 40 times and heads 60 times. The odds of this are low, but it could happen. Data collected by a poll is another way of finding experimental probabilities. A newspaper poll may find 50% of pet owners have a dog, 40% have a cat, and 10% have a bird or fish. The odds of a pet owner having a dog is 1 in 2.

Answer questions about the following table.

A poll was taken by a local newspaper about the color of cars in a small town.

1. What are the odds that a person in this town owns a blue car? A white car?

2. What color car is least likely to be owned by a town member? What are the odds that someone will have a car of that color?

Color of car	Number of owners
black	52
blue	48
brown	25
green	12
red	58
silver	30
white	52
other	60

© 2006 Walch Publishing

Independent Events

Independent events are those occurrences that happen that are not connected to one another. For example, the odds of you getting hit by lightning and winning the lottery on the same day are incredibly small. However, the two events are not really connected in any meaningful way. To find the probability of two independent events, you can multiply the odds of the first event by the odds of the second event.

Complete the following problems.

1. What are the odds of you rolling a 4 on a 6-sided die and then drawing the 5 of clubs from a deck of 52 playing cards?

2. What are the odds of you pulling a red marble from a sack of 10 red marbles and 15 blue marbles, and then flipping a quarter and having it come up heads?

3. What are the odds that a person who has a birthday on March 20 on a non-leap year rolls a die and gets a 6?

4. What are the odds of a person rolling a 4 on a 6-sided die, drawing the 5 of clubs from a deck of 52 playing cards, pulling a red marble from a sack of 10 red marbles and 15 blue marbles, flipping a quarter and having it come up heads, having a birthday on March 20 on a non-leap year, and rolling a die and getting a 6?

141

© 2006 Walch Publishing

Dependent Events

Dependent events are occurrences that happen in which the outcome of one event affects the outcome of another event. For example, if you were to put 3 peppermints, 3 chocolates, and 3 gumballs into a sack and then draw out three things, what would be the odds of getting a peppermint, a chocolate, and a gumball in that order, assuming the objects are not put back in once drawn out? There is a $\frac{3}{9}$ chance of getting a peppermint. If you draw a peppermint, the odds of getting a chocolate are now $\frac{3}{8}$ because there is one less item in the bag from the first drawing. If you draw a peppermint and then a chocolate, the odds are now $\frac{3}{7}$ because there are two items missing from the original nine. The total odds are $\frac{3}{9} \times \frac{3}{8} \times \frac{3}{7} = \frac{27}{504}$, which reduces to $\frac{3}{56}$ or about 5.4%.

A bag has 10 pennies, 5 nickels, 12 dimes, and 16 quarters. After a coin is drawn, it is not returned. (At the beginning of each question, however, all coins are back in the bag.) Answer the following questions.

1. What are the odds of drawing out 3 pennies?

2. What are the odds of drawing out 1 quarter, 1 dime, 1 penny, and 1 nickel?

142

© 2006 Walch Publishing

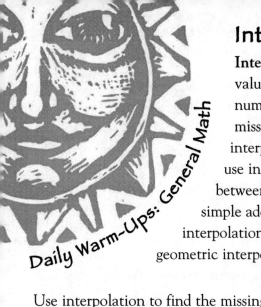

Interpolation

Interpolation is the process of using existing data and estimating a value or values between points in the set. For example, in the number sequence 1, 2, 3, 4, __, 6, 7, it is fairly obvious that the missing number is 5. The process of finding this number is interpolation. The prefix *inter-* means "among" or "between." We use interpolation when we are looking for missing numbers that are between numbers we already know. If the answer can be found by simple addition and subtraction, the process is sometimes called linear interpolation. If multiplication or division is used, the process is called geometric interpolation.

Use interpolation to find the missing numbers below.

1. 2, 4, 6, 8, _____, 12, 14

2. 2, 4, 8, 16, 32, ___, 128, 256

3. 100, 110, 120, 130, 140, _____, 160

4. $\frac{1}{2}, \frac{1}{4}, \frac{1}{8}, \underline{\hspace{1cm}}, \frac{1}{32}, \frac{1}{64}, \frac{1}{128}$

5. 321, 329, _____, 345, 353, 361, 369, _____, 385, 393

143

© 2006 Walch Publishing

Extrapolation

Extrapolation is the process of using existing data and estimating a value or values that go beyond the points in a set. For example, in the number sequence 3, 6, 9, 12, 15, 18, 21, ___, ___, you can probably tell that the missing numbers are 24 and 27. The process of finding these numbers is extrapolation. The prefix *extra-* means "beyond" or "outside." We use extrapolation when we are looking for missing numbers that are beyond numbers we already know. If the answer can be found by simple addition and subtraction, the process is sometimes called linear extrapolation. If multiplication or division is used, the process is called geometric extrapolation. Extrapolation can be tricky because it is not always clear if a set of numbers will behave the same outside of the data we know for sure.

144

Use extrapolation to find the missing numbers below.

1. 90, 95, 100, 105, _____, _____

2. 3, 6, 12, 24, 48, 96, _____, _____

3. $\frac{1}{2}$, $\frac{2}{6}$, $\frac{4}{9}$, $\frac{8}{27}$, $\frac{16}{81}$, _____, _____

4. $\frac{1}{3}$, $-\frac{3}{6}$, $\frac{9}{12}$, $-\frac{27}{24}$, $\frac{81}{48}$, _____, _____

5. 78, 72, 66, 60, 54, 48, _____, _____

© 2006 Walch Publishing

Survey Data

When you have a question or questions about a group of people, you can collect answers to those questions using a **survey.** The survey is usually in the form of a questionnaire that the people fill out and then return to you. Once you have the answers, you can organize the data you've collected to best determine how to understand it. A good survey will either be designed to test a random sample of people, or will be given to a very specific group. The questions should be designed in a way that allows the person taking the survey to answer the questions without any bias. For example, if one of your questions said "How do you feel about that jerk, President Jones, who is known to make small children cry?" you might expect that some people who do not already have a strong opinion might form one from reading your question.

Write five questions that you could use to survey your classmates or friends. Try to make your questions free from bias.

145

© 2006 Walch Publishing

Algebra

Algebra is a branch of mathematics that deals with math in terms of symbols. The symbols include a mix of numbers and letters and a variety of operators. Operators are the symbols that represent various mathematical functions, such as $+$, $-$, $\times$, $\div$, $\sqrt{}$, the ^{2} in 3^2, and so forth. Algebra also simplifies some of the operators, especially multiplication. The letter x is very common in algebra and is easily confused with the multiplication sign $\times$. The multiplication sign is often replaced with a $\bullet$ or is left out completely. For example, 3×4 can be written as $3 \bullet 4$ or $(3)(4)$. An algebra term such as $4 \times x$ might be better written as $4 \bullet x$ or simply $4x$.

Daily Warm-Ups: General Math

Explain the meaning of each term below.

1. $5x$

2. $7 \bullet 12$

3. xy

4. xy^2

5. $m^3 \bullet n$

146

© 2006 Walch Publishing

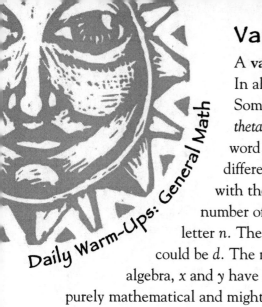

Variables

A **variable** is a symbol that is usually used to represent a number. In algebra, the variables are usually letters of the alphabet. Sometimes variables are letters from the Greek alphabet such as *theta* (Θ), which can be used to represent angle measures. The word *vary* means "to change," and variables can have many different values. It is often useful to choose a variable that starts with the letter of the value you are trying to find. For example, the number of students who like classical music could be represented by the letter n. The height of a building could be h. The distance to the moon could be d. The mass of an object could be m. When choosing variables in algebra, x and y have become two letters that are used often in expressions that are purely mathematical and might only represent numbers.

Choose a variable for each of the following quantities.

1. time
2. velocity
3. age
4. total number of marbles in a bag
5. length
6. volume

147

© 2006 Walch Publishing

Expressions

Algebraic expressions are combinations of numbers and variables. For example, if your height is represented as h, then twice your height would be $2h$. The term $2h$ is an expression. Algebraic expressions also must contain one mathematical operation. Below are some examples of algebraic expressions.

$$5 + x \qquad 5x$$
$$\frac{x}{8} \qquad 3x^2$$
$$y - 4 \qquad \frac{1}{7}y^3 + 4$$

Write an algebraic expression for each problem below.

1. 4 times the distance

2. 4 times the distance plus 5 feet

3. one twentieth of an unknown number

4. earning $20 an hour for an unknown number of hours

5. one half of an object's mass times its velocity squared

148

© 2006 Walch Publishing

Writing Equations

An **equation** is a mathematical statement that contains an equal sign (=). An algebraic equation is a mathematical statement that contains the equal sign as well as algebraic expressions. For example, the statement $x + 1 = 5$ is an algebraic equation.

Algebraic equations help us determine what information we know, and what information we are trying to find. For example, if you bought some items at a bake sale, you could find the cost of an item if you knew the total and the cost of the other items. If your cookies, a pie, and a cake cost $12 and the cookies and pie cost $9, then the cake cost $3. The algebraic expression could be $\$9 + c = \$12$, where c = cake.

Write an algebraic equation for each of the following.

1. Your height is Billy's height plus 13 inches.

2. The total cost of a meal is $3 per hamburger plus $2 per milkshake.

3. The total length of a road race is half the distance of your last race minus 500 meters.

4. The total cost for a group of people to go to the movies is $10.50 per person.

5. The total cost of filling the gas tank in a car is 14 gallons times the price of gasoline.

149

© 2006 Walch Publishing

Solving Equations

Some algebraic equations can be solved using simple mental math. For example, if you were given the problem $x + 2 = 5$, then you would be able to guess that the answer is 3 without really doing any math on paper or with a calculator. You are already doing algebra when you solve a problem like this, but you may not realize it. You might have subtracted the 2 from the 5, or maybe you tried different values for x until you found the one that worked, but you found the value of the variable that made the equation true.

Solve each algebraic equation below using mental math.

1. $m + 6 = 10$

2. $x^2 = 9$

3. $3x = 6$

4. $4t + 1 = 13$

5. $2x + 2x = 12$

6. $x^2 - 1 = 99$

150

© 2006 Walch Publishing

Distributive Property

When you add two numbers together, such as in the expression (3 + 2), the numbers 3 and 2 are addends. The **Distributive Property** says that to multiply the sum of a number, you multiply each addend by the number outside the parentheses. This is like saying $4(3 + 2) = 4(3) + 4(2)$. The same property holds true for algebraic expressions. The expression $3(x - 4) = 3(x) - 3(4) = 3x - 12$.

Simplify each expression below using the Distributive Property.

1. $4(x + 3)$

2. $10(4 - x)$

3. $x(x + 1)$

4. $6(x^2 + x)$

5. $8(25 + 6)$

6. $y^3(4 - x)$

151

© 2006 Walch Publishing

Algebra Properties

Here are some basic properties of math that can be applied to algebra.

Additive Identity: The sum of any number and zero is the number.

$$5 + 0 = 5 \qquad\qquad 1000 + 0 = 1000$$

Multiplicative Identity: The product of any number and 1 is the number.

$$10 \times 1 = 10 \qquad\qquad 138 \times 1 = 138$$

Commutative Property: The order that any numbers are added or multiplied does not change the sum or product.

$$1 + 3 = 3 + 1 \qquad\qquad 5 \times 2 = 2 \times 5$$

Associative Property: The way that numbers are grouped when they are added or multiplied does not change the sum or product.

$$(1 \times 2) \times 5 = 1 \times (2 \times 5) \qquad\qquad 3 + (2 + 6) = (3 + 2) + 6$$

Identify which of the four properties is being used in each example below.

1. $24 \times 3 = 3 \times 24$
2. $93 \times 1 = 93$
3. $12 + 0 = 12$
4. $1 \times x = x$
5. $x^2 + 3 = 3 + x^2$
6. $z + (3 + r) = (z + 3) + r$

152

© 2006 Walch Publishing

Solving Addition Equations

The equation $x + 3 = 8$ is a simple addition equation. The answer is 5, but how do you find the answer for a problem in which the answer is not so simple? The **Subtraction Property of Equality** states that when you subtract the same number from both sides of an equation, the two sides still have the same value. If you subtract a 3 from $x + 3$, you get x. If you subtract 3 from 8, you get 5. You already knew that the answer was 5, and now you have proved it.

This process also holds true for variables and expressions. For example, you can solve the problem below by subtracting x^2 from both sides of the equation.

$$10 + x^2 = x^2 + y$$

The answer is $y = 10$.

Solve the following problems using the Subtraction Property of Equality.

1. $12 + x = 15$

2. $y + 6 = 51$

3. $x + 1 = 10$

4. $9 + a = -400$

5. $d + 15 = -5$

6. $6 = 21 + z$

7. $16 = x + 83$

8. $1423 = r + 1214$

153

© 2006 Walch Publishing

Solving Subtraction Equations

The equation $x - 2 = 5$ is a simple subtraction equation. The answer is 7, but how do you find the solution for a problem in which the answer is not so simple? The **Addition Property of Equality** states that when you add the same number to both sides of an equation, the two sides still have the same value. If you add a 2 to $x - 2$, you get x. If you add 2 to 5, you get 7. You already knew that the answer was 7, and now you have proved it. This process also holds true for variables and expressions. For example, you can solve the problem below by adding x^3 to both sides of the equation.

$$15 - x^3 = y - x^3$$

The answer is $y = 15$.

Solve the following problems using the Addition Property of Equality.

1. $x - 102 = 15$

2. $y - 6 = 54$

3. $x - 34 = 10$

4. $a - 9 = -150$

5. $d - 130 = -5$

6. $11 = z - 28$

7. $17 = x - 93$

8. $1813 = r - 1144$

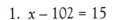

154

© 2006 Walch Publishing

Solving Multiplication Equations

The equation $2x = 6$ is a simple multiplication equation. The answer is 3, but how do you find the answer for a problem in which the answer is not so simple? The **Division Property of Equality** states that when you divide both sides of an equation by the same number, the two sides still have the same value. If you divide $2x$ by 2, you get x. If you divide 6 by 2, you get 3. You already knew that the answer was 3, and now you have proved it. This process also holds true for variables and expressions. For example, you can solve the problem below by dividing both sides of the equation by x.

$$12x = xy$$

The answer is $y = 12$.

Solve the following problems using the Division Property of Equality.

1. $4x = 12$

2. $3y = -27$

3. $45 = 9z$

4. $1.5r = 7.5$

5. $20x = -20$

6. $315 = 9t$

7. $49{,}875 = 25b$

8. $78x = 5928$

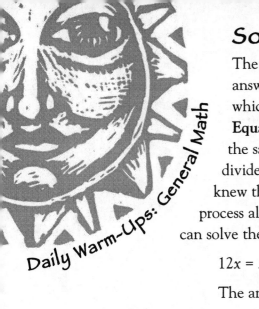

155

© 2006 Walch Publishing

Solving Division Equations

The equation $\frac{x}{4} = 2$ is a simple division equation. The answer is 8, but how do you find the solution for a problem in which the answer is not so simple? The **Multiplication Property of Equality** states that when you multiply both sides of an equation by the same number, the two sides still have the same value. If you multiply $\frac{x}{4}$ by 4, you get x. If you multiply 2 by 4, you get 8. You already knew that the answer was 8, and now you have proved it. This process also holds true for variables and expressions. For example, you can solve the problem below by multiplying both sides of the equation by x.

$$\frac{7}{x} = \frac{y}{x}$$

The answer is $y = 7$.

Solve the following problems using the Multiplication Property of Equality.

156

1. $\frac{x}{4} = 32$

2. $\frac{x}{11} = 715$

3. $\frac{r}{25} = 2$

4. $\frac{p}{-9} = -954$

5. $8.4 = \frac{t}{55}$

6. $6 = \frac{m}{8}$

7. $6.5 = \frac{x}{20}$

8. $-6 = \frac{w}{6}$

© 2006 Walch Publishing

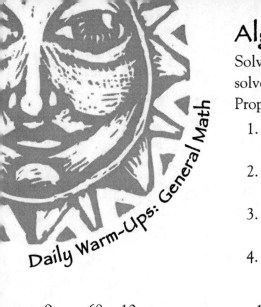

Algebra Equations Practice

Solve each problem below, and tell what property you used to solve it: Addition, Subtraction, Multiplication, or Division Property of Equality.

1. $27 + x = 88$

2. $22 + a = -250$

3. $x - 92 = 18$

4. $\frac{x}{11} = 15$

5. $y - 9 = 34$

6. $\frac{p}{-9} = -354$

7. $4x = 16$

8. $x + 14 = 10$

9. $x - 68 = 10$

10. $3y = -81$

11. $81 = 9z$

12. $1.5r = 9$

13. $\frac{x}{4} = 64$

14. $a - 20 = -150$

15. $\frac{r}{25} = 62$

16. $y + 9 = 61$

157

© 2006 Walch Publishing

Solving Two-Step Equations

Two-step equations contain two operations. Any combination of the Addition, Subtraction, Multiplication, and Division Properties of Equality may be needed to solve the problem. Generally addition and subtraction are done first before multiplication and division. For example, in the equation $3x + 1 = 7$, you would start by using the Subtraction Property of Equality and subtract 1 from both sides. The equation would then read $3x = 6$. You could then use the Division Property of Equality and divide both sides by 3. The final answer is $x = 2$.

Solve each two-step problem below.

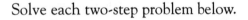

1. $5x + 4 = 14$

2. $2x - 8 = 32$

3. $40 = 3y + 1$

4. $7a - 12 = 9$

5. $100 = 5d - 25$

6. $80 + 3m = 50$

7. $500 + 2h = 300$

8. $431 - 8b = 399$

158

© 2006 Walch Publishing

Writing Two-Step Equations

A two-step equation combines two kinds of operations. A word problem that requires a two-step equation must be read carefully. For example, look at the parts of the following statement: Two less than three times a number is 25.

Three times a number means $3x$.

Two less than three times a number means $3x - 2$.

Two less than three times a number is 25 means $3x - 2 = 25$.

You can solve this problem by adding 2 to both sides and dividing both sides by 3. The answer is $x = 9$.

Write a two-step equation for each statement, and then solve.

1. Four more than four times a number is 24.

2. Half of a number plus 6 equals 10.

3. A number squared plus 10 equals 19.

4. 21 is the sum of 6 times a number plus 9.

5. Fifteen times a number plus 1 equals 46.

6. A number divided by 5 is added to 12 to get 15.

© 2006 Walch Publishing

Variables on Both Sides

Sometimes in algebra you will be faced with an equation that has variables on both sides of the equal sign. These variables can be combined or eliminated by using the same algebraic properties that you have used on simpler problems. Try to get all of the variables on one side of the equation and all of the numbers on the other. For example, in the equation $6x + 45 = 11x$, you need to collect the x terms together. If you subtract $6x$ from both sides, the equation becomes $45 = 5x$. Dividing both sides by 5 produces the answer $x = 9$.

Solve each equation below.

1. $2x + 24 = 8x$

2. $7y = 30 + 2y$

3. $2z + 50 = 6z + 10$

4. $24m + 50 = 100 - m$

5. $8r - 30 = 2r - 36$

6. $78 + x = 4x + 12$

160

© 2006 Walch Publishing

Inequalities

An **inequality** is a way of writing an equation that does not have to have an equal sign. The symbol < means less than. The symbol > means greater than. For example, 6 < 9 means that 6 is less than 9. Like equalities, inequalities must be true. The symbol ≤ means a number is less than or equal to. The symbol ≥ means greater than or equal to. For example, you must be 18 or older to vote. If the voting age is v, then the inequality statement for voting age would be $v \geq 18$.

Write an inequality statement for each of the following.

1. You must be over 4 feet tall to go on this ride.

2. You have to be 65 or older to collect Social Security.

3. Six-foot-tall Mark is taller than five-foot-tall Lily.

4. You must be 12 or younger to enter this contest.

5. The speed limit on the highway is between 45 and 65 mph.

161

© 2006 Walch Publishing

Solving Inequalities by Adding or Subtracting

Inequalities can contain variables, just as regular equations do. You can solve an inequality using the same properties of addition and subtraction that allow you to solve a regular equation. For example, for the inequality $x + 10 > 30$, you can subtract 10 from both sides and get $x > 20$. If the equality was $x - 6 < 10$, you could add 6 to both sides and get $x < 16$. This also holds true for inequalities using the less than or equal to symbol ($\leq$) and the greater than or equal to symbol ($\geq$).

Solve the following inequalities.

1. $x + 4 < 12$

2. $y - 8 > 3$

3. $z + 3 < 4$

4. $25 + x > 2x$

5. $-21 - 2z \geq z$

6. $y - 1 \leq 8$

162

© 2006 Walch Publishing

Solving Inequalities by Multiplying or Dividing

You can solve inequalities by multiplying or dividing just as with regular algebraic equalities. There is one small difference, however. When you multiply or divide by negative numbers, you must reverse the direction the equality symbol points for the statement to remain true. For example, in the inequality $25 - 5x < 60$, you would subtract 25 from both sides to get $-5x < 35$. Then you would divide both sides by -5 to get $x > -7$. *Note:* The < symbol is now a > symbol. If the symbol had not been reversed, the answer would have been $x < -7$. If you tried -8 in the original problem, the answer would be $65 < 60$, which is not true.

Daily Warm-Ups: General Math

Solve the following inequalities.

1. $3x + 4 < 19$

2. $4y - 8 > 4$

3. $2z + 3 < 21$

4. $24 + 6x > 2x$

5. $5z - 20 \geq z$

6. $6y - 1 \leq 35$

163

© 2006 Walch Publishing

Functions

In mathematics, a **function** is a relationship in which one thing depends on another. Generally in a function, an input number is chosen and put into the function. Solving the function produces output. The function is usually one or more calculations that take the number you put in and transform it into an answer. A function is often written as $f(x)$. This is read "the function of x," or "f of x." For example, look at the function $f(x) = 2x + 1$. The table below summarizes this concept.

Input (x)	Function (2x + 1)	Output f(x)
1	2(1) + 1	3
2	2(2) + 1	5
3	2(3) + 1	7
4	2(4) + 1	9
5	2(5) + 1	11

Complete the table below.

Input (x)	Function (3x + 2)	Output f(x)
0		
1		
2		
3		
4		

164

© 2006 Walch Publishing

Graphing Linear Functions

Using a function to generate output will lead to the production of a set of ordered pairs. You may remember that we use ordered pairs when plotting points on the coordinate plane. For example, look at the chart below. The output, $f(x)$, can be used as the y coordinate.

Input (x)	Function (2x + 1)	Output f(x)
1	2(1) + 1	3
2	2(2) + 1	5
3	2(3) + 1	7
4	2(4) + 1	9
5	2(5) + 1	11

The ordered pairs created are (1, 3), (2, 5), (3, 7), (4, 9), and (5, 11). When you graph these points, you will see that they all lie on the same straight line. You can connect these points to see all the other points on the line as well.

Graph the line generated by each function below. Use x values from at least 0 to 5.

1. $f(x) = 2x - 2$

2. $f(x) = x + 1$

165

© 2006 Walch Publishing

Slope from Points

The **slope** of a line is a measure of how steep the line is. A line that is completely flat has no incline and therefore has a slope of zero. The steeper a line gets, the greater the value of the slope. For a line on the coordinate plane, the slope is negative if the value of y decreases as the value of x increases. The slope is positive if the value of y increases as the value of x increases. The slope formula is $m = \frac{y_2 - y_1}{x_2 - x_1}$ in which m is the slope. Any two points from the line are represented by the ordered pairs (x_1, y_1) and (x_2, y_2).

Find the slope of each line using the ordered pair provided.

1. $(5, 5)$, $(3, 3)$

2. $(25, 1)$, $(22, 8)$

3. $(8, 9)$, $(12, 20)$

4. $(-2, 6)$, $(-10, 8)$

5. $(-10, 4)$, $(-1, 6)$

6. $(2, 2)$, $(-6, -6)$

166

© 2006 Walch Publishing

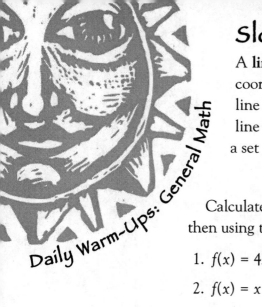

Slope from Functions

A **linear function** can be used to generate ordered pairs of coordinates. Those coordinates all fall on the same line, and that line has a slope. There is more than one way to find the slope of a line from the function. One way is to use the function to generate a set of ordered pairs and then use the slope formula $m = \frac{y_2 - y_1}{x_2 - x_1}$.

Calculate the slope of each line below by generating ordered pairs and then using the slope formula.

1. $f(x) = 4x - 2$

2. $f(x) = x + 1$

3. $f(x) = 3x - 3$

4. $f(x) = -3x + 6$

5. $f(x) = 8x - 3$

6. $f(x) = 2x + 10$

167

© 2006 Walch Publishing

Slope-Intercept Form

Functions are usually written with $f(x)$ to the left of the equal sign. When using the coordinate plane, $f(x)$ is usually used for the y value and is written as such in the equation. The function $f(x) = 3x + 1$ is more conveniently written as $y = 3x + 1$. This form of a line equation is called the **slope-intercept form.** The slope-intercept form of a line is written as $y = mx + b$, where m is the slope of the line and b is the value at which the line crosses (intercepts) the y-axis. When a line crosses the y-axis, the x value is always zero. This is a useful tool for finding the slope and location of a line on the coordinate plane. Not all equations are written in this form, but can be rearranged algebraically.

168

Put each equation below into slope-intercept form.

1. $5x + y = 14$

2. $2x - y = 32$

3. $40 = 3x + y$

4. $7x - y = 9$

5. $y = 5x - 25$

6. $y + 3x = 50$

7. $y + 2x = 300$

8. $y - 8x = 399$

© 2006 Walch Publishing

Scatter Plots

A **scatter plot** is a kind of graph that can be used to find the relationship between two variables. Scatter plots consist of a number of points graphed from ordered pairs. If the points come close to lying on a straight line, the two variables are related. The two scatter plots below are examples.

a 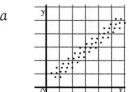   *b*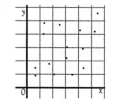

Scatter plot a shows a clear relationship between *x* and *y*. As *x* increases, *y* increases. Scatter plot b shows no clear relationship between *x* and *y*.

Make a scatter plot that shows the current height of people in your class versus their month of birth. Does the plot reveal a pattern or not?

169

© 2006 Walch Publishing

Graphing a System of Equations

A **system of equations** is a set of two or more equations. These equations may be linear in nature, and you can graph them. Any place where the lines cross one another on the coordinate plane is a place where the lines have identical coordinate points (x, y). These identical coordinates are called the solution to the system of equations. If there is no solution, then the lines do not cross and must be parallel. If two lines are identical and go through all the same points, then all the coordinates are solutions.

Find the solution to each system of equations below by graphing on a separate sheet of graph paper.

170

1. $y = x$ and $y = -x + 4$

2. $y = x + 4$ and $y = -2x - 5$

3. $y = 3x - 3$ and $y = -3x + 3$

4. $y = -\frac{1}{2}x + 4$ and $y = x + 1$

© 2006 Walch Publishing

Graphing Linear Inequalities

To graph a linear inequality, you need to first graph the line of the inequality as if it were an equality. This means that you would take an inequality such as $y < x + 3$ and graph it as if it were the line $y = x + 3$. Once you have graphed the line, choose a test point to see if the equality is true. $(0, 0)$ is a good point to try because the simplification is usually easy. For example, the equality $y < x + 3$ with $(0, 0)$ substituted becomes $0 < 3$. This is true, so on your graph you would shade in the side of the line where the point $(0, 0)$ lies. If it were not true, you would shade the portion of the graph on the other side of the line.

Graph each inequality below on a separate sheet of graph paper.

1. $y < 2x + 5$

2. $y > 3x - 6$

3. $y < -2x + 3$

4. $y > -x + 4$

171

© 2006 Walch Publishing

Simplifying Algebraic Expressions

The simplest form of an algebraic expression has no parentheses and no like terms. In the expression $3x + 5 + 2x = 12$, the $3x$ and $2x$ are like terms. These like terms can be simplified by combining them into one term, $5x$. An expression such as $x(5 +1)$ can be simplified into $5x + x$. That expression can be further simplified into $6x$. The single term $6x$ is the completely simplified version of the original expression.

Simplify each algebraic expression below.

1. $4(x +3)$

2. $-2(x + 6)$

3. $6x + 3 - 2x + 8$

4. $-y + 5 + 9 - 8y$

5. $7m + 3m$

6. $10r - 3r$

7. $6(3x + 4 + 2x - 6 + 5x)$

8. $(5)(4z - 5 - 5z - 8 - 12z + 30)$

Daily Warm-Ups: General Math

172

© 2006 Walch Publishing

Linear and Nonlinear Functions

A **linear function** is a function that has a graph that is a straight line. A **nonlinear function** is a function that produces a graph that is not a straight line. Such functions produce curved lines and do not represent a constant rate of change. One simple way of checking whether a function is linear or nonlinear is to see if you can rewrite the equation in slope-intercept form, $y = mx + b$. If you cannot put an equation into this form, it must be nonlinear. For example, if you try to solve the equation $y(x + 10) = 4$ for y, you would get $\frac{4}{x + 10}$. This is not in slope-intercept form, so the graph of this equation would not be linear.

Determine whether each function below is linear or nonlinear.

1. $2y - 8 = 6x$

2. $2xy = 12$

3. $3y - 3x = 18$

4. $4(y - 1) = 8x$

5. $3y - 12 = 9x^2$

6. $\frac{4y}{x - 2} = 4x$

173

© 2006 Walch Publishing

Daily Warm-Ups: General Math

Graphing Quadratic Functions

A **quadratic function** is a function in which the highest possible power is 2. That means, for example, that the equation $y = 3x + 1$ is not a quadratic function. The equation $y = 3x^2 + 1$ is a quadratic function because the x is squared. To graph a quadratic function, it is usually best to create a table of x and y values and then plot the points in order. Using the values −3, −2, −1, 0, 1, 2, 3 is common and often a good place to start. You may need to plot more points for the pattern to appear, but for simple quadratic graphs, these points should produce smooth curves with an obvious pattern.

Graph each quadratic function below on a separate sheet of graph paper.

1. $y = x^2$

2. $y = 2x^2 + 1$

3. $y = -x^2 - 1$

4. $y = x^2 - 4$

174

© 2006 Walch Publishing

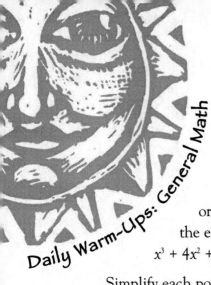

Simplifying Polynomials

The various terms in an expression are called **monomials.** For example, x^2, 4, $-y^2$, $6t$, and $-3z^2$ are all monomials. When groups of these monomials are connected by addition and subtraction, they are called **polynomials.** For example, $x + 1$, $2x + 5$, and $3x^2 + 2x - 4$ are polynomials. Remember that the simplest form of an algebraic expression is one that has no parentheses and no like terms. When polynomials contain monomials of different powers, you generally order the monomials by decreasing powers when possible. For example, the expression $2x + 4x^2 + 5 + x^3$ would be more appropriately written as $x^3 + 4x^2 + 2x + 5$.

Simplify each polynomial below.

1. $6a + 5b + 6 + 4a + 9b$

2. $5x^2 - x + 12 + 5x + 3x^2$

3. $3(2x + 1) + 3x^2 + 2x$

4. $8z + 4y^2 + 9z^2 - 3y^2 + 2z^2$

5. $3r + 6r + 8r - 5r + r + 4r^2$

6. $11 - 35x + 12x - 6x^2 + 14$

175

© 2006 Walch Publishing

Adding Polynomials

Polynomials can be added by collecting and adding their like terms. For example, the two polynomials below can be broken into similar terms and then added.

$3x^2 + 4x - 12$ and $5x^2 + 3x + 4$ can be rewritten as

$(3x^2 + 5x^2) + (4x + 3x) + (-12 + 4)$, which equals $8x^2 + 7x - 8$.

Add the polynomials below.

1. $(x + 1) + (2x + 4)$

2. $(5y + 1) + (2y - 4)$

3. $(2x^2 + 3) + (6x^2 + 8)$

4. $(4x^2 + 4) + (-2x^2 - 1)$

5. $(2x^2 + 5x + 1) + (6x^2 + 4x + 8)$

6. $(8x^2 + 2x + 2) + (-12x^2 - 5x - 21)$

7. $(-4z^2 + 3z + 12) + (-3z^2 - 6z + 8)$

8. $(8x^3 + 2x + 4x^2 + 7 + 4x) +$
$(16 + 4x + 3x^3 + 8x - 9x^2)$

176

© 2006 Walch Publishing

Subtracting Polynomials

Polynomials can be subtracted by collecting and subtracting their like terms. One other way to do this is to use the **additive inverse.** For example, the additive inverse of the second polynomial below can be found by multiplying all the terms in the polynomial by –1. Once this is done, you can use addition as when adding polynomials.

$(7x + 12) - (3x + 4)$ can be rewritten as

$(7x + 12) + (-3x - 4)$, which is then rewritten as

$(7x - 3x) + (12 - 4)$, which equals $4x + 8$.

Subtract the polynomials below.

1. $(5x + 7) - (2x + 4)$

2. $(2y + 8) - (2y - 4)$

3. $(6x^2 + 2) - (6x^2 + 8)$

4. $(3x^2 + 5) - (-2x^2 - 1)$

5. $(8x^2 + 4x + 1) - (6x^2 + 4x + 8)$

6. $(3x^2 + 3x + 2) - (-12x^2 - 5x - 21)$

7. $(-6z^2 + 2z + 12) - (-3z^2 - 6z + 8)$

8. $(9x^3 + 6x + 4x^2 + 13 + 4x) -$
   $(26 + 4x + 2x^3 + 8x - 10x^2)$

177

© 2006 Walch Publishing

Multiplying and Dividing Monomials

Remember that to multiply two powers that have the same base, you add their exponents. For example, $x^2 \cdot x^5 = x^7$. To divide two powers that have the same base, you subtract their exponents. For example, $x^8 \div x^3 = x^5$. If the monomials contain numbers, you multiply or divide them separately from the powers. For example, $3x^2 \cdot 4x^5 = 12x^7$. The same is true for division. For example, $16x^9 \div 2x^5 = 8x^4$.

Solve the following problems.

1. $x^2 \cdot x^3$

2. $z^4 \cdot z^8$

3. $y^{14} \div y^4$

4. $x^{13} \div x^7$

5. $2x^4 \cdot 2x^5$

6. $3y^3 \cdot 4y^4$

7. $12x^8 \div 2x^3$

8. $100y^{10} \div 20y^4$

178

© 2006 Walch Publishing

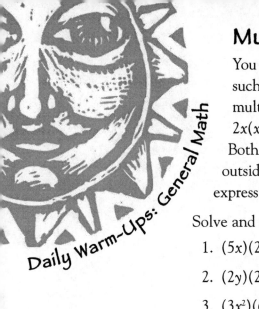

Multiplying Monomials and Polynomials

You can use the Distributive Property to simplify an expression such as $3(x + 2)$ into $3x + 6$. You can also use this property to multiply monomials and polynomials. For example, the expression $2x(x + 4)$ can be simplified as $(2x)(x) + (2x)(4) = 2x^2 + 8x$. *Note: Both terms in the binomial were multiplied by the monomial outside of the parentheses. This also simplifies the algebraic expression by eliminating the parentheses.*

Solve and simplify the following problems.

1. $(5x)(2x + 4)$

2. $(2y)(2y - 4)$

3. $(3x^2)(6x^2 + 8)$

4. $(5x)(-2x^2 - 1)$

5. $(4x)(6x^2 + 4x + 8)$

6. $(3x)(-12x^2 - 5x - 21)$

7. $(-2z^2)(-3z^2 - 6z + 8)$

8. $(4x)(26 + 4x + 2x^3 + 8x - 10x^2)$

179

© 2006 Walch Publishing

Multiplying Binomials and Binomials

The FOIL method is a way to help you remember the steps when you multiply two binomials. Here is an example of the FOIL method for multiplying the two binomials $(x + 3)$ and $(2x + 4)$.

1. Multiply the FIRST terms of each binomial: $(x)(2x) = 2x^2$

2. Multiply the OUTER terms: $(x)(4) = 4x$

3. Multiply the INNER terms: $(3)(2x) = 6x$

4. Multiply the LAST terms: $(3)(4) = 12$

Finally, you can add all the terms together, making the answer $2x^2 + 10x + 12$. *Note:* $4x$ and $6x$ were like terms and were collected together.

Solve the following problems using the FOIL method.

1. $(5x + 7)(x + 4)$

2. $(y + 8)(y - 4)$

3. $(6x + 2)(6x^2 + 8)$

4. $(3x + 5)(-2x - 1)$

5. $(8x^2 + 4x)(6x^2 + 8)$

6. $(3x + 2)(-5x - 21)$

7. $(-6z^2 + 12)(-3z^2 + 8)$

8. $(9x^3 + 13)(4x - 10)$

180

© 2006 Walch Publishing

1. 1. –20 2. 20 3. 300 4. –300
   5.
      –5   –2   0 1 2   8 9

   6.
      –150 –100 –75   0 25 50   150

2. 1. 8
   2. 56
   3. 22
   4. 6
   5. 100
   6. 524
   7. 66
   8. 30
   9. 99
   10. 5

3. 1. 1, 1, 1, 1, 1, 2, 2, 2, 2, 3, 3, 3, 4, 5, 6, 6, 7, 8
   2. 17, 17, 17, 17, 17, 18, 18, 18, 19, 19, 19, 20, 20, 21, 21, 21
   3. –300, –100, –75, –25, 0, 50, 150, 175, 200
   4. –33, –30, –21, –16, –12, –8, –7, –2

   5. –341, –306, – 45, 2, 55, 654, 921, 1000

4. 1. 17
   2. 3
   3. –9
   4. –16
   5. 85
   6. –35
   7. –195
   8. –305

5. 1. 3
   2. 17
   3. –21
   4. –8
   5. –35
   6. 85
   7. –305
   8. –195

6. 1. 64
   2. –40
   3. 30
   4. 99
   5. –176

6. –625
7. 5250
8. –12,500

7. 1. 40
2. –40
3. 40
4. 6
5. –6
6. 9
7. 334
8. –58

8. 1. prime
2. composite 1, 3, 9
3. composite, 1, 2, 3, 4, 6, 12
4. neither
5. composite, 1, 5, 7, 35
6. composite, 1, 2, 4, 8, 16
7. composite, 1, 2, 4, 8
8. composite 1, 2, 5, 10, 25, 50
9. prime
10. neither
11. neither

12. composite, 1, 2, 5, 10
13. composite 1, 3, 5, 15
14. prime

9. 1. 9^4
2. h^7
3. m^2n^5
4. 10^3
5. $2^2, 3^3$
6. 625
7. 64
8. 4^5

10. Negative values of all the following are also correct.
1. 2
2. 11
3. 4
4. 7
5. 6
6. 7.07
7. 9.22
8. 70.71
9. 1000

Daily Warm-Ups: General Math

10. 16.4
11. 1. 74
2. 116
3. 115
4. 1530
12. 1. whole number
2. integer
3. rational number
4. whole number
5. rational number
6. rational number
7. integer
8. rational number
13. 1. rational
2. irrational
3. rational
4. rational
5. rational
6. irrational
7. irrational
8. irrational
14. 1. 9

2. 16
3. 8
4. 50
15. 1. 0.000
2. 12.35
3. 0.04
4. 1.01
5. 75.8
6. 5.1
7. $16.88
8. $7.00
9. $8.20
10. $1599.30
16. 1. 20.285
2. 0.028
3. 15.69
4. 27.9
5. 123.232594
6. 18.887
17. 1. 7.11
2. 5.29
3. 0.00022

4. 2.67
5. 9.99
6. 3.33
7. 61.998
8. 19.75
18. 1. 7.32
2. 0.012
3. 18.64
4. 1.032
5. 14.664
6. 4.635
19. 1. 1.0302
2. 0.0000444
3. 27.56
4. 96.032002
5. 0.00001
6. 0.000669
7. 47.795
8. 0.1375
20. 1. 5.4
2. 0.00025
3. 5.08

4. 0.284
5. 0.0017
6. 1.161
21. 1. 4.99861
2. 0.00024
3. 4.35714
4. 0.27711
5. 1.54545
6. 0.91677
22. 1. 4
2. 25
3. 2
4. 100
5. 15
6. 75
7. 11
8. 2
23. 1. 40
2. 8
3. 21
4. 72
5. 72

6. 360
7. 210
8. 180

24.   1. 7/27
     2. 3/5
     3. 1/4
     4. 9/10
     5. 1/2
     6. 1/5

25.   1. 4 2/3
     2. 5 1/3
     3. 5 1/4
     4. 7 1/7
     5. 41 1/2
     6. 6 2/7

26.   1. 19/3
     2. 42/5
     3. 32/3
     4. 29/4
     5. 146/7
     6. 4/3
     7. 37/16

     8. 131/11

27.   1. 2/3, 3/4
     2. 3/7, 3/5
     3. 2/8, 3/9
     4. 5/27, 2/9, 1/3
     5. 1/8, 1/4, 17/64
     6. 17/80, 32/100, 9/25, 2/5

28.   1. 3/25
     2. 3/20
     3. 9/100
     4. 29/50
     5. 6/25
     6. 1/8
     7. 33/200
     8. 1/10
     9. 19/20
     10. 16/25

29.   1. 0.67
     2. 0.75
     3. 0.60
     4. 0.43
     5. 0.33

6. 0.22
7. 0.19
8. 0.25
9. 0.13
10. 0.27
11. 0.36
12. 0.21

30. 1. 2 1/3
2. 1/2
3. 1
4. –5/7
5. 2
6. 4/9
7. 2 1/27
8. –17 1/4
9. 7/8
10. 1/8
11. 22/25
12. –9/40

31. 1. 1 11/12
2. 11/20
3. 14/15

4. –29/63
5. –7 19/36
6. 61/104
7. 13/320
8. 1 197/200

32. 1. 5/6
2. 3/20
3. –1/5
4. 8/21
5. 5/12
6. –1/18
7. 1 17/18
8. –3/52
9. 17/320
10. 1 17/200

33. 1. 8/15
2. 3 3/4
3. 1 4/5
4. –27/56
5. 4/15
6. 5 5/9
7. 13/48

8. –1 21/64
34. 1. 4 1/3
2. 6 2/7
3. 10 1/5
4. 2 7/8
5. 11 19/20
6. 31 2/7
35. 1. –2
2. 2
3. –3 1/5
4. 3/8
5. 1 2/5
6. 7 1/4
7. 2/7
8. 1 1/6
36. 1. 4 1/12
2. 10 1/2
3. –5 2/5
4. 1 17/64
5. 3 3/10
6. –2 38/49
7. 2 1/10

8. 5 7/9
37. 1. 1/3
2. 1 13/14
3. –2 2/5
4. 1
5. 2 7/24
6. –2 1/8
7. 1 1/14
8. 1 3/13
38. 1. 3 1/3
2. 1 12/35
3. 3 1/5
4. 2 1/8
5. 2 19/20
6. 1 17/21
7. 1 29/35
8. 3 22/45
39. 1. 3
2. 33/35
3. 2
4. 5/8
5. 2 11/20

6. 10/21
7. 34/35
8. 1 32/45

40.   1. 7/15
2. 1/3
3. 2 3/5
4. 1 1/24
5. 1/4
6. 16/21
7. 4/7
8. 4 20/27

41.   1. 2 11/12
2. 3/49
3. 4 1/16
4. 3/8
5. 6 1/4
6. 7/12
7. 3 4/7
8. 1/6

42.   2 and 3 are correct.
1 and 4 are incorrect.
5. 6

6. 56
7. 12
8. 300

43.   1. 0.125
2. 0.28
3. 0.16
4. 0.21
5. 0.8
6. 14.3%
7. 8%
8. 35%
9. 90%
10. 40%

44.   1. 25%
2. 75%
3. 66%
4. 115%
5. 89%
6. 0.12
7. 0.61
8. 0.125
9. 0.152

10. 0.9999
45.   1. 1.35
     2. 15
     3. 4.8
     4. 2.33
     5. 3/2
     6. 15/5
     7. 7/4
     8. 5/4
     9. 333%
    10. 109%
    11. 350%
    12. 466%
46.   1. 0.0005
     2. 0.0025
     3. 0.00033
     4. 0.0099
     5. 0.0012
     6. 0.2%
     7. 0.15%
     8. 0.33%
     9. 0.68%

10. 0.09%
47.   1. 43.8
     2. 45
     3. 297
     4. 4.95
     5. 655.38
     6. 3.92
     7. 13.31
     8. 2.652
     9. 53.452
    10. 10
    11. 11
    12. 278.69242
48.   1. 46.7%
     2. 20
     3. 9.1
     4. 10.83
     5. 16.65
     6. 116.14
49.   1. 21.76
     2. 48.4%
     3. 24

4. 19.375
5. 72%
6. 612.5
7. 0.34
8. 34.4%
9. 4.05
10. 79.8
11. 173%
12. 15.94

50. 1. 35%
2. 86.7%
3. 23.6%
4. 775%

51. 1. $1.00
2. $2.50
3. $0.21
4. $1.53
5. $163.51
6. $23.62
7. $49.44
8. $32.91

52. 1. $108

2. $36.40
3. $48.75
4. $2010
5. $1000
6. $4.50
7. $7.20
8. $10.17

53. 1. $15
2. $75
3. $2400
4. $1210
5. $145,000
6. $25.60

54. 1. $196.72
2. $11,592.74
3. $110,401.98
4. $453.32
5. $1344.56

55. The measurement system is the same so that communication of data can be standardized.

56. 1. m or cm
2. cm or mm

3. m
4. km
5. m or cm
6. mm
7. mm
8. m

57. 1. l
     2. ml
     3. ml
     4. l or ml
     5. l
     6. ml
     7. l
     8. l
     9. ml
     10. ml

58. 1. kg
     2. g
     3. kg
     4. mg
     5. kg
     6. mg

7. g
8. kg or g

59. 1. 4 feet
     2. 12 yards
     3. 21,120 yards
     4. 31,680 feet
     5. 63,360 inches
     6. 20.83 yards
     7. 1953 feet
     8. 270 inches
     9. 2.84 miles
     10. 14.59 miles
     11. 4188 inches
     12. 1.41 miles

60. 1. 2.125 cups
     2. 3.44 pints
     3. 8 quarts
     4. 128 pints
     5. 24 cups
     6. 96 cups
     7. 208 ounces
     8. 1536 ounces

9. 72 cups
10. 40 quarts
61. 1. 192 ounces
   2. 7000 pounds
   3. 51,200 ounces
   4. 22.4 short tons
   5. 31.25 long tons
   6. 7616 pounds
   7. 160 ounces
   8. 4906.25 pounds
   9. 26.5 long tons
  10. 7840 pounds
62. 1. 1277.5 days
   2. 23,301.6 hours
   3. 720 minutes
   4. 4320 seconds
   5. 1.01 days
   6. 12.57 years
   7. 2,312,640 minutes
   8. 2,520,000 seconds
   9. 31,536,000 seconds
  10. 92,160 minutes

63. 1. 5.28 pounds
   2. 7.27 kilograms
   3. 24.948 liters
   4. 14.76 gallons
   5. 2217.75 milliliters
   6. 294.22 fluid ounces
   7. 40 kilometers
   8. 218.75 miles
   9. 30.48 centimeters
  10. 288.58 inches
64. 1. 7.5×10^4
   2. 1.3×10^6
   3. 5.5×10^2
   4. 8×10^{12}
   5. 9.5×10^{-9}
   6. 73,000
   7. 2,000,000
   8. 366,000,000
   9. 120
  10. 0.000001
65. 1. 3
   2. 3

3. 1
4. 4
5. 2
6. 10
7. 4
8. 6
66. Answers will vary. Sample answers:
   1. 0.1 mile
   2. 5 mph
   3. 1 pound
   4. 0.01 gallons
   5. 0.1 degree
   6. 1 mm
   7. 1/16 inch
   8. 1/16 pound
67. Answers will vary.
68. Answers will vary.
69. 1. eye colors in my class
   2. brown; speckled
   3. 3.350%; 5.0%
   4. Answers will vary.
70. 1. 10 meters; 22.5 meters

   2. 7 seconds; 10 seconds
   3. 25 seconds, which is a guess
71. 1. February; June and July
   2. summer; fall and winter
   3. 2.9%; 11.8%
72. 1. frequency chart
   2. line graph
   3. circle graph
   4. bar graph
73. 1. $\begin{bmatrix} 3 & 7 \\ 14 & 22 \end{bmatrix}$

   2. $\begin{bmatrix} 12 & 17 & 7 \\ 9 & 20 & 5 \\ 7 & 23 & 35 \end{bmatrix}$

   3. cannot be added

   4. $\begin{bmatrix} 3 & 3 & 2 \\ -6 & -14 & -6 \\ 2 & 3 & 4 \end{bmatrix}$

74. 1. 24
   2. 28
   3. 12
   4. 25

5. 24
6. 24
75. 1. 9
    2. 18
    3. 24
    4. 12
    5. 16
    6. 40
76. 1. right
    2. acute
    3. obtuse
    4. right
    5. straight
    6. acute
    7. obtuse
    8. right
77. All angles are adjacent.
    1. complementary
    2. supplementary
    3. adjacent
    4. supplementary
    5. adjacent

    6. complementary
78. 1. a. 45°
      b. 135°
      c. 45°
    2. d. 40°
      e. 140°
79. Answers will need to be checked with a protractor.
80. a. 55°
    b. 125°
    c. 45°
    d. 125°
    e. 55°
    f. 125°
81. Answers will vary.
82. a. 6
    b. 5
    c. 10
    d. 4
    e. 60°
    f. 5
    g. 110°
    h. 70°

83. 1. multiple answers
    2. multiple answers
    3. not symmetrical
    4. vertical line through center
    5. multiple answers
    6. not symmetrical
    7. vertical line through center
    8. vertical line through center
84. 1, 4, 5, 6, and 7 are similar.
85. 1. $x = 24$
    2. $a = 16$, $b = 18$
    3. $x = 21$, $y = 14$
86. 1, 5: congruent
    2, 6: neither
    3, 4: similar
87. Answers will vary.
88. Answers must be checked by measurement.
89. Answers must be checked by measurement.
90. 1. 50°
    2. 65°
    3. 90°
    4. 48°

5. 45°
6. 90°
91. 1. 1.41
    2. 5
    3. 11.66
    4. 8.54
    5. 6.63
    6. 22.45
92. 1. $a = 6$, $b = 8.49$
    2. $c = 8$, $d = 6.93$
    3. $e = 17.68$, $f = 17.68$
    4. $g = 15$, $h = 25.98$
93. 1. 72 cm²
    2. 24 cm²
    3. 30 cm²
    4. 48 cm²
    5. 56.7 cm²
    6. 120 cm²
94. 1. 12 in²
    2. 12 in²
    3. 10 in²
95. 1. 21 cm²

Daily Warm-Ups: General Math

2. 32.5 in^2

3. 55.5 cm^2

4. 42.5 cm^2

5. 18 in^2

6. 35 in^2

96. 1. 31.4 cm

    2. 25.12 cm

    3. 18.84 cm

    4. 37.68 in

    5. 41.448 cm

    6. 33.284 in

97. Choice c is the best answer.

98. 1. 200 cm^2

    2. 54 cm^2

    3. 97 in^2

    4. 126 in^2

    5. 242.5 cm^2

99. 1. 6 faces, 12 edges, 8 vertices

    2. 5 faces, 9 edges, 6 vertices

    3. 6 faces, 12 edges, 8 vertices

    4. 5 faces, 8 edges, 5 vertices

100. 1. Objects could include cans, pipes, logs, and so forth.

    2. Objects could include party hats, tepees, and so forth.

    3. Objects could include tennis balls, ball bearings, and so forth.

101. Answers will vary.

102. 1. 216 cm^3

    2. 192 in^3

    3. 108 cm^3

    4. 600 in^3

103. 1. 810 cm^3

    2. 612.5 in^3

    3. 80 cm^3

104. 1. 254.34 cm^3

    2. 339.12 in^3

    3. 2769.48 cm^3

    4. 2355 in^3

    5. 6028.8 cm^3

105. 1. 480 cm^3

    2. 616 in^3

    3. 170.7 cm^3

106. 1. 267.9 cm^3

      2. 98.5 in³
      3. 6427.8 cm³
107. 1. 113.04 in³
      2. 267.9 cm³
      3. 904.32 cm³
      4. 17,148.6 in³
      5. 8.18 cm³
108. 1. 152 cm²
      2. 384 in²
      3. 450 cm²
109. 1. 664.8 cm²
      2. 777.7 in²
      3. 206 cm²
110. 1. 226.08 cm²
      2. 339.12 in²
      3. 1099 cm²
111. 1. 84 cm²
      2. 84.8 in²
112. 1. 251.2 in²
      2. 1130.4 cm²
      3. 1073.88 in²
113. 1. 113.04 in²

      2. 200.96 cm²
      3. 452.16 cm²
      4. 3215.36 in²
      5. 19.625 cm²
114. Answers will vary.
115. 1. 4
      2. 12.5
      3. 298.8
      4. 12
      5. 7
      6. 77
      7. 95.075
      8. 9
      9. 48.6
      10. 152
116. 1. 5
      2. 9
      3. 284
      4. 25
      5. 50
      6. 20
      7. 89.5

8. 4
117. 1. 1
2. 5
3. 123
4. 8
5. 17

118.

Stem	Leaf
1	77788899
2	0011111123
3	1344459
4	0455
5	14
6	7
7	03

119. 1. 1, 3, 3, <u>3</u>, 4, 4, 4, <u>4</u>, 5, 5, 6, <u>7</u>, 7, 8, 8
2. 22, 23, <u>23</u>, 23, 24, 24, 30, <u>33</u>, 35, 35, 40, <u>41</u>, 44, 45, 50
3. 152, 155, 155, <u>156</u>, 158, 164, 166, <u>169</u>, 171, 172, 174, <u>175</u>, 177, 178, 180
4. 1.7, 1.7, <u>1.7</u>, 1.8, 1.8, 1.8, <u>1.9</u>, 1.9, 1.9, 2.0, <u>2.0</u>, 2.1, 2.1, 2.1

120. 1. 4
2. 4
3. 19
4. 4
5. 0.19

121. 1. −3, 13
2. 126.5, 202.5
3. −0.4, 6.8

122. 1.

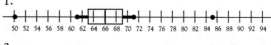

2.

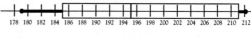

123.

1. 4
2. 2.5
3. 4
4. − 4
5. 36, 42, 48
6. 88, 77, 66

7. 182, 224, 266
8. 21, 35, 49

124. 1. 0.5
2. –3
3. 128, 512, 2048
4. 702, 768

125.

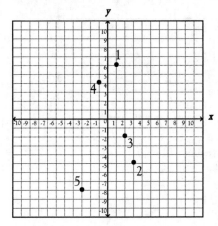

126. 1. 2.83
2. 7

3. 11.7
4. 8.2
5. 9.2
6. 11.3

127. The marked points should now be as follows:
1. (–5, –3)
2. (6, 5)
3. (8, 9)
4. (9, –4)

128.

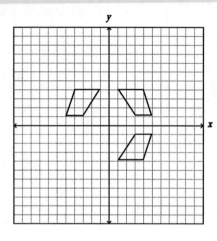

129.

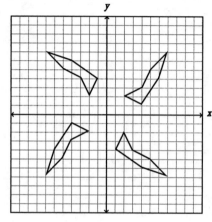

130.   1. 1:5
       2. Answers will vary.
       3. 3:4
       4. 1:6
       5. 100:1
       6. 36:1

131.   1. Answers will vary. Sample answer: 55 mph
       2. a. 4 batteries for $3.00

   b. $3.50 for a dozen eggs
   c. $2.99 for a six-pack of soda
   d. They have the same value.
132. 1. 1/6
      2. 12/35
      3. Answers will vary.
133. 1. 26
      2. 12
      3. 52
      4. 475
134.

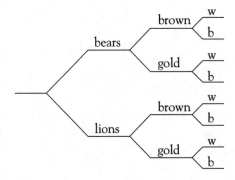

135. 1. Answers will vary.
      2. Answers will vary.
      3. 1/16
      4. Answers will vary.
136. 1. 216
      2. 2704
      3. 1,757,600
      4. 4,569,760,000
137. 1. 132,600
      2. 39,270
      3. 720
138. 1. 10
      2. 10
139. 1. 1/36
      2. 1/52
      3. 1/2652
      4. 1/16
140. 1. 48/337; 52/337
      2. green; 12/337
141. 1. 1/312
      2. 1/5
      3. 1/2190

4. 1 = 3, 416, 400

142. 1. 120/12,341
2. 40/12,341

143. 1. 10
2. 64
3. 150
4. 1/16
5. 337, 377

144. 1. 110, 115
2. 192, 384
3. 32/243, 64/729
4. −243/96, 729/192
5. 42, 36

145. Answers will vary.

146. 1. 5 times x
2. 7 times 12
3. x times y
4. x times y times y
5. m times m times m times n

147. 1. t
2. v
3. a

4. m
5. l
6. V

148. 1. $4d$
2. $4d + 5$
3. $x/20$
4. $\$20x$
5. $1/2 \ mv^2$

149. 1. $h = b + 13$
2. $t = \$3h + \$2m$
3. $t = 1/2 \ d - 500$
4. $t = \$10.50p$
5. $t = 14p$

150. 1. 4
2. 3 or −3
3. 2
4. 3
5. 3
6. 10 or −10

151. 1. $4x + 12$
2. $40 - 10x$
3. $x^2 + x$

4. $6x^2 + 6x$
5. $150 + 48$
6. $4y^3 - xy^3$

152. 1. CP
2. MI
3. AI
4. MI
5. CP
6. AP

153. 1. 3
2. 45
3. 9
4. − 409
5. −20
6. −15
7. −67
8. 209

154. 1. 117
2. 60
3. 44
4. −141
5. 125

6. 39
7. 110
8. 2957

155. 1. 3
2. −9
3. 5
4. 5
5. −1
6. 35
7. 1995
8. 76

156. 1. 128
2. 7865
3. 50
4. 8586
5. 462
6. 48
7. 130
8. −36

157. 1. 61, Subtraction
2. −272, Subtraction
3. 110, Addition

4. 165, Multiplication
5. 43, Addition
6. 3186, Multiplication
7. 4, Division
8. – 4, Subtraction
9. 78, Addition
10. –27, Division
11. 9, Division
12. 6, Division
13. 256, Multiplication
14. –130, Addition
15. 1550, Multiplication
16. 52, Subtraction

158.   1. 2
2. 20
3. 13
4. 3
5. 25
6. –10
7. –100
8. 4

159.   1. $4x + 4 = 24$, $x = 5$

2. $x/2 + 6 = 10$, $x = 8$
3. $x^2 + 10 = 19$, $x = 3$ or -3
4. $6x + 9 = 21$, $x = 2$
5. $15x + 1 = 46$, $x = 15$
6. $x/5 + 12 = 15$, $x = 15$

160.   1. 4
2. 6
3. 10
4. 2
5. 1
6. 22

161.   1. $h > 4$
2. $a \geq 65$
3. $6 > 5$, or M > L
4. $a \leq 12$
5. $45 \leq s \leq 65$

162.   1. $x < 8$
2. $y > 11$
3. $z < 1$
4. $25 > x$
5. $-7 \geq z$
6. $y \leq 9$

163. 1. $x < 5$
      2. $y > 3$
      3. $z < 9$
      4. $-6 < x$
      5. $5 \leq z$
      6. $y \leq 6$

164.

Input (x)	Function ($3x + 2$)	Output f(x)
0	3(0) + 2	2
1	3(1) + 2	5
2	3(2) + 2	8
3	3(3) + 2	11
4	3(4) + 2	14

165.

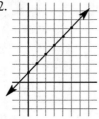

1.   2.

166. 1. 1

          2. $-7/3$
          3. $11/4$
          4. $-1/4$
          5. $2/9$
          6. 1

167. 1. 4
      2. 1
      3. 3
      4. -3
      5. 8
      6. 2

168. 1. $y = -5x + 14$
      2. $y = 2x - 32$
      3. $y = -3x + 40$
      4. $y = 7x - 9$
      5. $y = 5x - 25$
      6. $y = -3x + 50$
      7. $y = -2x + 300$
      8. $y = 8x + 399$

169. Answers will vary.

170. 1. (2, 2)
      2. (−3, 1)

3. (1, 0)
4. (2, 3)

171.   1.
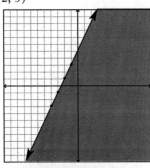

2.
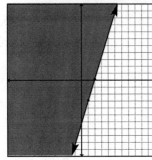

3.
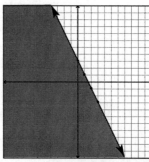

4.
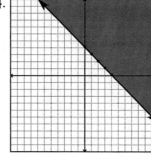

172.   1. $4x + 12$
2. $-2x - 12$

3. $4x + 11$
4. $-9y + 14$
5. $10m$
6. $7r$
7. $60x - 12$
8. $-65z + 85$

173. 1. linear
2. nonlinear
3. linear
4. linear
5. nonlinear
6. nonlinear

174. 1.

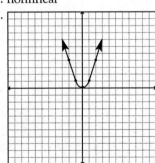

2.

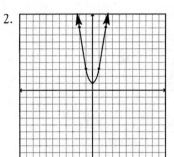

3.

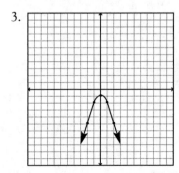

4.

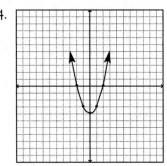

7. $-7z^2 - 3z + 20$
8. $11x^3 - 5x^2 + 18x + 23$

177. 1. $3x - 3$
2. 12
3. -6
4. $5x^2 + 6$
5. $2x^2 - 7$
6. $15x^2 + 8x + 23$
7. $-3z^2 + 8z + 4$
8. $7x^3 + 14x^2 - 2x - 13$

178. 1. x^5
2. z^{12}
3. y^{10}
4. x^6
5. $4x^9$
6. $12y^7$
7. $6x^5$
8. $5y^6$

175. 1. $10a + 14b + 6$
2. $8x^2 + 4x + 12$
3. $3x^2 + 8x + 3$
4. $y^2 + 11z^2 + 8z$
5. $4r^2 + 13r$
6. $-6x^2 - 23x + 25$

176. 1. $3x + 5$
2. $7y - 3$
3. $8x^2 + 11$
4. $2x^2 + 3$
5. $8x^2 + 9x + 9$
6. $-4x^2 - 3x - 19$

179. 1. $10x^2 + 20x$
2. $4y^2 - 8y$
3. $18x^4 + 24x^2$
4. $-10x^3 - 5x$

5. $24x^3 + 16x^2 + 32x$
6. $-36x^3 - 15x^2 - 63x$
7. $6z^4 + 12z^3 - 16z^2$
8. $8x^4 - 40x^3 + 48x^2 + 104x$

180. 1. $5x^2 + 27x + 28$
2. $y^2 + 4y - 32$
3. $36x^3 + 12x^2 + 48x + 16$
4. $-6x^2 - 13x - 5$
5. $48x^4 + 24x^3 + 64x^2 + 32x$
6. $-15x^2 - 73x - 42$
7. $18z^4 - 84z^2 + 96$
8. $36x^4 - 90x^3 + 52x - 130$

Turn downtime into learning time!

For information on other titles in the

Daily *Warm-Ups* series,

visit our web site: walch.com